ANTHONY SLY obtained his degree at the University of Reading and has since taught at various Hertfordshire schools and colleges. At present head of the Mathematics Department at Queens' School, Bushey, Herts., he is also a mathematics examiner and author of *A Short History of Computing* (Hertfordshire County Council Advisory Unit for Computer-based Education). He is also the author of *Modern Mathematics Model Answers* in the Key Facts series.

GCE O-Level Passbooks

BIOLOGY, R. J. Whitaker, B.Sc. and J. M. Kelly, B.Sc., M.I.Biol.

CHEMISTRY, C. W. Lapham, M.Sc.

COMPUTER STUDIES, R. J. Bradley, B.Sc.

ECONOMICS, J. E. Waszek, B.Sc.(Econ.)

ENGLISH LANGUAGE, Robert L. Wilson, M.A.

FRENCH, G. Butler, B.A.

GEOGRAPHY, R. Knowles, M.A.

GEOGRAPHY, BRITISH ISLES, D. Bryant, B.A. and R. Knowles, M.A.

GERMAN, A. Nockels, M.A.

HISTORY, Political and Constitutional (1815-1951), L. E. James, B.A., M.Litt.

HISTORY, Social and Economic (1815-1951), M.C. James, B.A. and L. E. James, B.A., M.Litt.

HUMAN BIOLOGY, S. Cantle, B.Sc., M.Med.Sci.

MODERN MATHEMATICS, A. J. Sly, B.A.

PHYSICS, B. P. Brindle, B.Sc.

TECHNICAL DRAWING, P. J. Barnett, D.S.C., M.C.C.Ed., Adv.Dip.Ed.

GCE O-Level Passbook

Modern Mathematics

A. J. Sly, B.A.

Published by Charles Letts & Co Ltd
London, Edinburgh and New York

Diary House, Borough Road, London SE1 1DW
Reprinted 1986
Made and printed by Charles Letts (Scotland) Ltd
ISBN 0 85097 373 2

Contents

Introduction

The contents of this book take into account the syllabuses and recent examination papers of all relevant GCE O-level examining boards. A common-core syllabus is presented here, together with a great many minor topics which appear only on certain syllabuses. A considerable amount of this material is also relevant to the CSE examinations, which are often styled on O-level GCE.

Though theoretically it is incorrect to break mathematics up into different topics because they are all interdependent, it is necessary in a book of this nature to present the material under different section headings for the sake of order and clarity. In this book, the algebraic topics appear first, followed by geometry, arithmetic, statistics, trigonometry and calculus.

The style of mathematics examination papers varies between boards. Some set two evenly-balanced papers, whereas others set one paper of multiple-choice or short questions and one of longer, more searching questions. Furthermore, the style and emphasis of the questions varies between boards, so which ever board you are taking, you are advised to consult the past papers of the board concerned. (Addresses appear on page 192.) All the examples provided in this book are very closely modelled on actual examination questions.

An O-level pass in mathematics has always been high on the list of basic job requirements, especially for scientific careers. All scientific subjects require a mathematical background to at least O-level, so for some readers the O-level mathematics course will be just the beginning of their studies. More recently, mathematical techniques have been adopted in less closely-related subjects too; for example, statistics are used in biology, geography and economics.

Modern mathematics, or to give it a better title, the New Approach to mathematics, was introduced to bring about a more flexible attitude to the subject. The formal proofs and long calculations have been dropped in favour of material requiring more original thought. However, the fundamental arithmetic, algebra and geometry are included here because they are essential to any basic mathematical training.

Chapter 1
Sets

Language

A set is a well-defined collection of elements. Here are some examples of sets and the correct way to write them:

$A = \{0, 1, 2, 3, 4, 5, 6, 7, 8, 9\}$ or $A = \{\text{the digits}\}$ $V = \{a, e, i, o, u\}$
$N = \{1, 3, 5, 7 \ldots 21\}$ $E = \{0, 2, 4, 6 \ldots\}$

A capital letter and brackets { } are required when listing a set. The set N consists of the odd numbers from 1 to 21 inclusive. E is an **infinite** set. There are an infinite number of even numbers. $N = \{x: x \text{ is odd}, 1 \leqslant x \leqslant 21\}$ is another way of writing set N.

The following definitions are illustrated with these sets:

$\mathscr{E} = \{0,1,2,3 \ldots 9\}$ $A = \{2,3,4,5,6\}$ $B = \{3,4,5\}$ $C = \{5,3,4\}$
$D = \{4,5,6,8\}$ $F = \{7,8\}$

The **universal** set $\mathscr{E}$ denotes all elements under consideration. $6 \in A$ states that 6 is an **element** of A, $7 \notin B$ states that 7 is **not an element** of B.
Equal sets have the same elements, irrespective of order. $B = C$.
$B \subset A$ means that B is a **subset** of A. It can also be written $A \supset B$ which states that set A **contains** set B. See figure 1(a).
An **empty** or **null** set has no elements. It is written { } or $\varnothing$.
The **intersection** of sets A and D is the set made by the elements in set A **and** set D. It is written $A \cap D = \{4, 5, 6\}$. See figure 1(b).
The **union** of sets A and D is the set containing the elements in A **or** D **or both**. Each element is listed **once**. It is written $A \cup D = \{2, 3, 4, 5, 6, 8\}$. See figure 1(b).
The **complement** of the set A, written A', is the set containing the elements of $\mathscr{E}$ not in A. $A' = \{0, 1, 7, 8, 9\}$. See figure 1(c).
Disjoint sets have no common elements. A and F are such sets. $A \cap F = \varnothing$. See figure 1(d).
The **number of elements** in set D is 4. This is written $n(D) = 4$.
Venn diagrams are used to clarify set manipulation. The universal set is a rectangle; all other sets are balloons as in figure 1. No element occurs on the boundary. The first four diagrams demonstrate the definitions given above.

Verify the truth of the following using the sets $\mathscr{E}, A, B, C$:
$A \cup \mathscr{E} = \mathscr{E}$: $A \cap \mathscr{E} = A$: $A \cup A = A$: $A \cap A = A$
$C \subset A \therefore A \cup C = A$ and $A \cap C = C$

$\varnothing \subset A$. The empty set is a subset of every set.
$A \subset \mathscr{E}$, $B \subset \mathscr{E}$, etc. Every set is a subset of the universal set.
These statements are true for sets in general.

The laws of sets

The **commutative law** $A \cup B = B \cup A$ and $A \cap B = B \cap A$.
The **associative law**

for Union $(A \cup B) \cup C = A \cup (B \cup C)$.

for Intersection $(A \cap B) \cap C = A \cap (B \cap C)$.

Figure 1(e) shows $(A \cup B) \cup C$. First shade $(A \cup B)$ vertically. Then shade C horizontally. The total area shaded is $(A \cup B) \cup C$.
Figure 1(f) shows $A \cup (B \cup C)$. First shade $(B \cup C)$ vertically. Then shade A horizontally. The total area shaded is $A \cup (B \cup C)$.
The total area in each case is the same, so $(A \cup B) \cup C = A \cup (B \cup C)$.

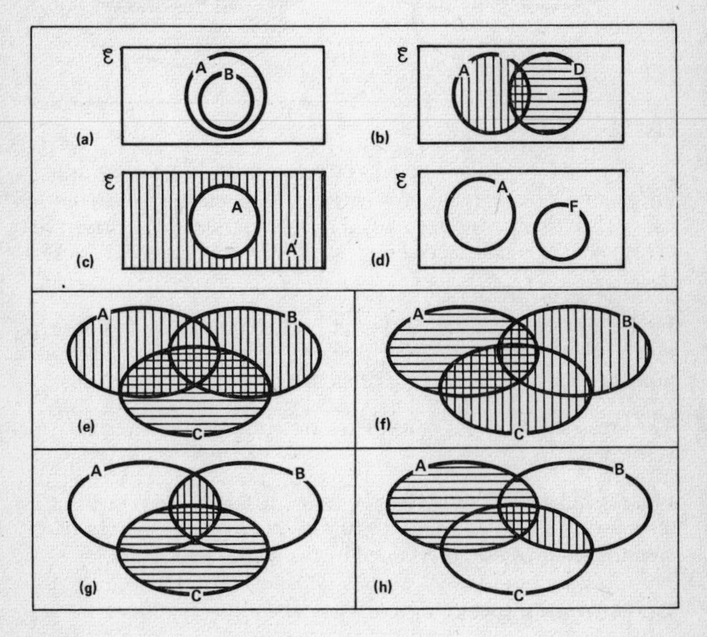

Figure 1

Verify that $(A \cap B) \cap C = A \cap (B \cap C)$ by using figures 1(g) and (h). Remember that the area shaded both ways is required for the intersection of sets.

The **distributive law**

Intersection over union: $A \cap (B \cup C) = (A \cap B) \cup (A \cap C)$
Union over intersection: $A \cup (B \cap C) = (A \cup B) \cap (A \cup C)$

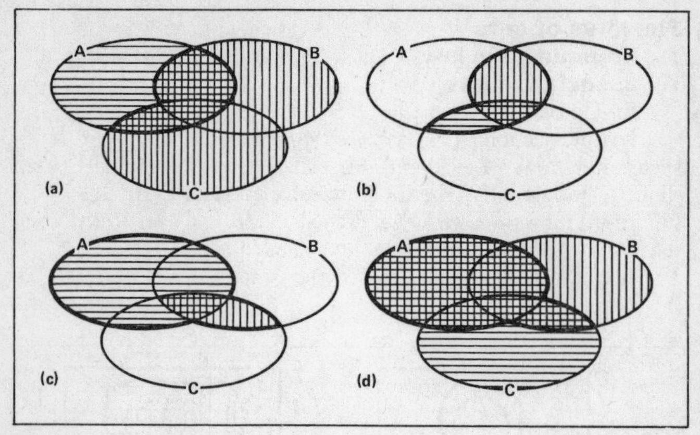

Figure 2

To show that $A \cap (B \cup C) = (A \cap B) \cup (A \cap C)$ see figures 2(a) and (b). Figure 2(a) represents $A \cap (B \cup C)$. First shade $(B \cup C)$ vertically. Then shade A horizontally. For the intersection of these two areas take the region shaded both ways. It is bounded by a thicker line.

Figure 2(b) represents $(A \cap B) \cup (A \cap C)$. First shade $(A \cap B)$ vertically. Then shade $(A \cap C)$ horizontally. For the union of these two take the total area shaded. It is bounded by a thicker line also.

The areas shown in (a) and (b) are the same so the statement is verified.

Figures 2(c) and (d) verify that $A \cup (B \cap C) = (A \cup B) \cap (A \cup C)$. (c) shows $A \cup (B \cap C)$ and (d) shows $(A \cup B) \cap (A \cup C)$. In each case the required area is enclosed by the thicker line.

De Morgan's laws: 1 $(A \cap B)' = A' \cup B'$
2 $(A \cup B)' = A' \cap B'$

To verify 1 see figures 3(a) and (b). In figure 3(a) $(A \cap B)'$ is the complement of $(A \cap B)$ i.e. all the shaded area of $\mathscr{E}$ except $(A \cap B)$. In figure 3(b) A' (all of $\mathscr{E}$ except A) is shaded vertically. In figure 3(b) B' (all of $\mathscr{E}$ except B) is shaded horizontally.

In both diagrams the total area shaded is the same so the statement is verified.

Now verify Law 2 using figures 3(c) and (d).

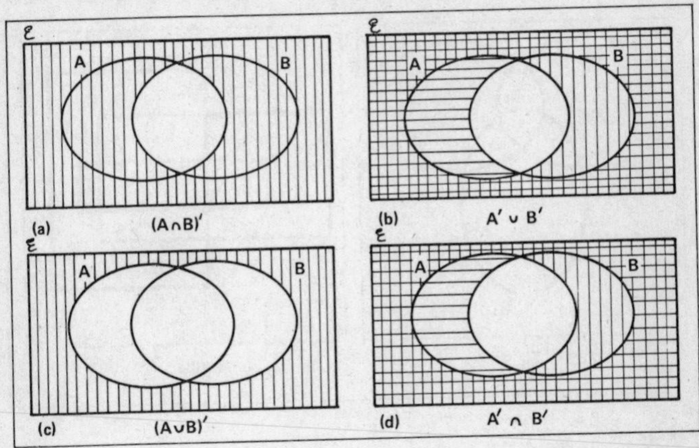

Figure 3

Example It is given that $\mathscr{E} = \{x; x \leqslant 15, x$ is a natural number (see page 28)$\}$, $A = \{$numbers divisible by 3$\}$, $B = \{5, 7, 12, 15\}$, $C = \{$prime numbers$\}$.

(a) List the following sets: (i) $A \cap B$ (ii) $(A \cup B)'$ (iii) $A \cap B \cap C$ (iv) $(A \cap B) \cup (A \cap C)$.

(b) What are (i) $n(A)$ and (ii) $n(A \cup B)'$?

The universal set consists of all the natural numbers less than or equal to 15. So

$$\mathscr{E} = \{1, 2, 3, 4, 5, 6, 7, 8, 9, 10, 11, 12, 13, 14, 15\}$$

The other sets are:

$A = \{3, 6, 9, 12, 15\}$ $\quad B = \{5, 7, 12, 15\}$ $\quad C = \{2, 3, 5, 7, 11, 13\}$

(a) (i) $A \cap B = \{12, 15\}$, those elements in A and B.

(ii) $A \cup B = \{3, 5, 6, 7, 9, 12, 15\}$, listing common elements once. So $(A \cup B)' = \{1, 2, 4, 8, 10, 11, 13, 14\}$, all of $\mathscr{E}$ not in $A \cup B$.

(iii) $A \cap B \cap C = \varnothing$; no elements are common to sets A, B and C.

(iv) $A \cap B = \{12, 15\}$ and $A \cap C = \{3\}$. The union of these two sets is required, $\therefore$ $(A \cap B) \cup (A \cap C) = \{3, 12, 15\}$.

Do not forget the brackets { } when listing sets.

(b) (i) $n(A) = 5$. There are five elements in set A.
(ii) $n(A \cup B)' = 8$. Count the elements in (a) (ii).

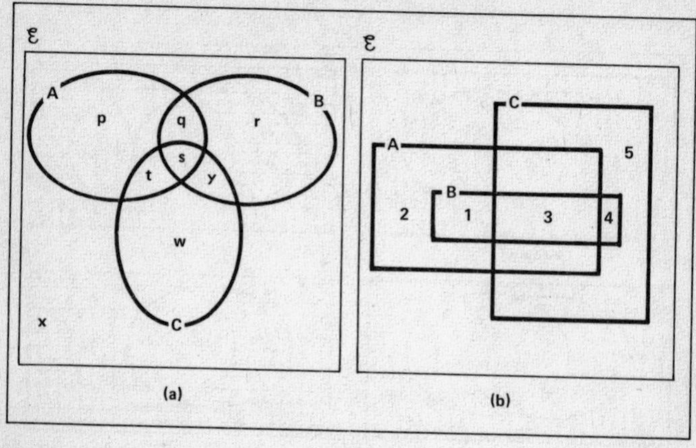

Figure 4

Example In figure 4(a) write in set notation the areas labelled by (i) p, (ii) q, (iii) x. What areas are represented by (iv) $A \cup (B \cap C)$ and (v) $(A \cup C) \cap B'$?

(i) p is the part of A not in B and not in C.

$$\therefore \text{the area } p = A \cap B' \cap C'.$$

(ii) The area q is the part of $A \cap B$ which is not in C.

$$\therefore q = (A \cap B) \cap C'.$$

(iii) The area x is the part of $\mathscr{E}$ which is not in A and not in B and not in C. $\therefore x = A' \cap B' \cap C'$. This can also be written $(A \cup B \cup C)'$, the extension of De Morgan's second law for three sets.

(iv) $B \cap C =$ areas s and y. A = areas p, q, s, t.

$$\therefore A \cup (B \cap C) = \text{areas } p, q, s, t, y.$$

(v) $A \cup C =$ areas p, q, t, s, y, w. Those in B are to be excluded.

$$\therefore (A \cup C) \cap B' = p, t, w.$$

In table 1, other areas in figure 4(a) are labelled. It is important to know such areas when solving problems using sets.

Area labelled	Set notation	Area labelled	Set notation
p, q, r, s, t, y	$A \cup B$	r	$A' \cap B \cap C'$
q, s	$A \cap B$	$t, s, y.$	$(A \cup B) \cap C$
s	$A \cap B \cap C$	$q, s, t, y, w.$	$(A \cap B) \cup C$
p, q, r	$(A \cup B) \cap C'$	$q, s, t.$	$A \cap (B \cup C)$

Table 1

Example In figure 4(b), which areas are represented by (i) $A \cap B \cap C$, (ii) $A \cap B' \cap C'$, (iii) $(A \cup B) \cap C$? Write in set notation (iv) area 1, (v) areas 4 and 5 together.

(i) The area common to A, B and C is 3, $\therefore$ $A \cap B \cap C =$ area 3.

(ii) The area which is in A and not in B and not in C is 2, $\therefore$ $A \cap B' \cap C' =$ area 2.

(iii) $(A \cup B) =$ the areas 2, 1, 3, 4. C is the areas 3, 4, 5, $\therefore$ $(A \cup B) \cap C =$ the areas 3 and 4.

(iv) Area 1 is in A and B but not in C. $\therefore$ area 1 $= (A \cap B) \cap C'$.

(v) Areas 4 and 5 are in the union of B and C but not in A, $\therefore$ the set notation is $(B \cup C) \cap A'$.

Example In a class of 35 pupils 27 study maths., 23 study physics, and 19 study chemistry; 18 study maths. and physics. 15 study maths. and chemistry while 13 study physics and chemistry; 10 pupils study maths., physics and chemistry. (i) How many pupils do not study maths. or physics or chemistry? (ii) How many study maths. and physics but not chemistry? (iii) How many study physics only?

'27 study maths.' does not mean that they study maths. only. The 27 include the physicists and chemists who also study maths. It simplifies the problem if this information is written out using set notation.

The universal set	$n(\mathscr{E}) = 35$
Those studying maths.	$n(M) = 27$
physics	$n(P) = 23$
chemistry	$n(C) = 19$
maths. and physics	$n(M \cap P) = 18$
maths. and chemistry	$n(M \cap C) = 15$
physics and chemistry	$n(P \cap C) = 13$
Those studying all three subjects	$n(M \cap P \cap C) = 10$

Draw a clear Venn diagram as in figure 5(a). Place the 10 in the space for $M \cap P \cap C$.

$n(M \cap P) = 18$ means that the overlap between M and P contains 18 pupils. 10 of these have already been accounted for, so the remainder of the overlap contains 8.

The 5 and 3 can be filled in using a similar process with $n(P \cap C)$ and $n(M \cap C)$.

The complete set M contains 27 pupils. $8+10+5 = 23$ of these have been accounted for, so the remaining space of M is $27-23 = 4$.

The remaining spaces of P and C can be filled in by a similar process using $n(P)$ and $n(C)$. Now in figure 5(b) all the spaces are filled and the questions can be answered.

(i) The number studying maths. or physics or chemistry = $n(M \cup P \cup C) = 4+8+2+3+10+5+1 = 33$, $\therefore$ the number of pupils who do not study maths. or physics or chemistry, $n(M \cup P \cup C)' = 35-33 = 2$.

(ii) Those studying maths. and physics and not chemistry = $n(M \cap P \cap C')$. From figure 5(b) the answer is 8.

(iii) Those studying physics only = $n(M' \cap P \cap C')$ which is 2.

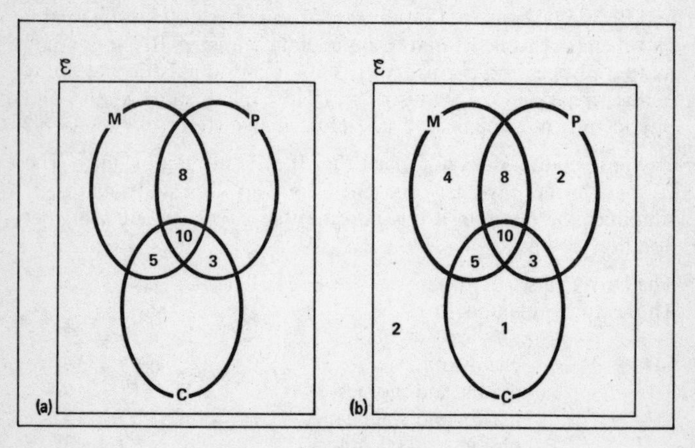

Figure 5

Example A survey of summer games enjoyed by 40 boys shows that 18 of them like cricket, 26 like tennis, 30 like swimming; 14 like cricket and swimming, 13 like tennis and cricket, 17 like tennis and swimming. All of the boys like at least one activity. Find: (i) how many like tennis, cricket and swimming; (ii) how many like swimming only; (iii) how many like tennis and swimming but not cricket.

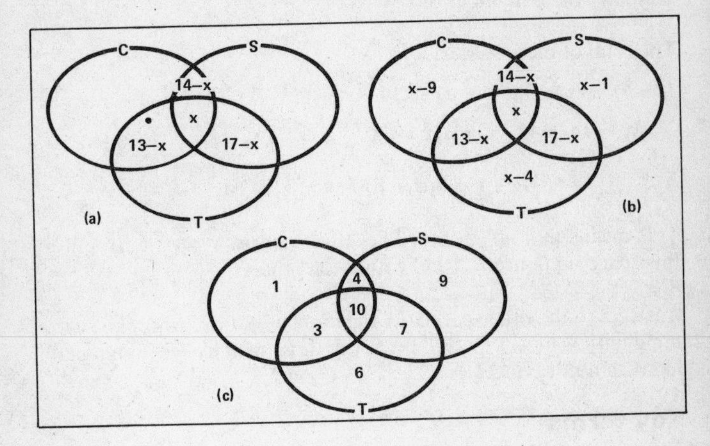

Figure 6

Express the details in set notation:

$$n(C) = 18 \qquad n(S) = 30 \qquad n(T) = 26$$
$$n(C \cap S) = 14 \qquad n(T \cap C) = 13 \qquad n(T \cap S) = 17$$

Draw a Venn diagram. There is no need to include a rectangle for the universal set because all of the boys like at least one activity. However, in this question no number can be definitely placed in any of the spaces, so let $x = n(C \cap S \cap T)$.

$n(C \cap S) = 14$ so in figure 6(a) place $14-x$ in the remainder of the overlap of C and S. Fill in the other spaces in a similar way.

In figure 6(b) the remaining spaces are filled in as follows. There are 18 boys in C.

$$\begin{aligned} \text{The number accounted for} &= (14-x) + x + (13-x) \\ &= 27 - 2x + x \\ &= 27 - x. \end{aligned}$$

$\therefore$ the remaining empty space in C is $18 - (27 - x)$

$$= 18 - 27 + x$$
$$= x - 9.$$

The number of boys in $S = 30$.
Number accounted for $= (14-x)+x+(17-x) = 31-x$
$\therefore$ the remainder $= 30-(31-x) = x-1$.
Similarly the remainder of $T = x-4$.

The total of these spaces = 40

$$(x-9) + (x-4) + (x-1) + (14-x) + (13-x) + (17-x) + x = 40.$$

$$\therefore 3x - 14 + 44 - 3x + x = 40$$

$$\therefore x + 30 = 40 \qquad \therefore x = 10$$

This means that $n(C \cap S \cap T) = 10$, the answer to part (i). Fill in this space on figure 7(c) and with it the others.

Answers to the other parts: (ii) the number of boys who like swimming only $= n(S \cap C' \cap T') = 9$; (iii) those who like swimming and tennis but not cricket $= n(T \cap S \cap C') = 7$.

Key terms

$\mathscr{E}$, the universal set of reference.
$5 \in A$, 5 is an element of the set A.
$24 \notin A$, 24 is not an element of the set A.
$B \subset A$ or $A \supset B$ both mean that B is a subset of A.
$\varnothing$ or $\{\}$ are both symbols for the empty set.
$A \cap B$ is the intersection of sets. Elements must be in both A and B.
$A \cup B$ is the union of sets. Elements must be in A or B or both.
A' is the complement of the set A. Those elements of $\mathscr{E}$ which are not in A.

A knowledge of the laws of sets, although useful, is not essential for GCE at O-level. The laws of sets are often included as problems in GCE O-level and CSE papers and are included here mainly as worked examples.

Chapter 2
Basic Algebra

Negative or directed numbers

$$-\quad -6\quad -5\quad -4\quad -3\quad -2\quad -1\quad 0\quad 1\quad 2\quad 3\quad 4\quad 5\quad 6\quad +$$

To simplify $-(-3)$ on the number line, start at 0, look in the negative direction and move 3 units backwards. The final position is $+3$. Hence $-(-3) = +3$. Similarly $+(-3) = -3$ and $-(+3) = -3$.

Verify that $(-2)+(-3) = -2-3 = -5$ and
$(+4)-(-6) = 4+6 = 10$.

Multiplication and division

$(+3) \times (+2) = +6 \qquad (+6) \div (+2) = +3$
$(+3) \times (-2) = -6 \qquad (+6) \div (-2) = -3$
$(-3) \times (+2) = -6 \qquad (-6) \div (+2) = -3$
$(-3) \times (-2) = +6 \qquad (-6) \div (-2) = +3$

To show that $(-3) \times (-2) = +6$, imagine a vehicle which passes a point 0 and moving backwards at 3 m/s (-3 m/s). 2 seconds ago (-2) it would have been 6 m in the $+$ direction, i.e. $+6$ m.

$(+2)\times(-3) = -6 \Leftrightarrow (-6)\div(-3) = +2$, division being the inverse of multiplication. The other rules for division are shown similarly.

A common error is to say that 'two minuses make a plus'. $(-2)\times(-3) = +6$ but $(-2)+(-3) = -5$ is not plus! The rule is that under $\times$ or $\div$ two **unlike** signs give a $-$ answer and two **like** signs give a $+$ answer.

The laws of indices

An **index** is a **power**. $x^4 = x \times x \times x \times x$ is called 'x to the power 4'. x is called the **base**.

$$x^3 \times x^2 = x \times x \times x \times x \times x = x^5 \text{ or } x^{3+2}$$

$$x^3 \div x^2 = \frac{x \times x \times x}{x \times x} = x^1 \text{ or } x \text{ or } x^{3-2}$$

$$(x^3)^2 = x^3 \times x^3 = x^6 \text{ or } x^{3\times 2}$$

When the bases are the same: to **multiply** numbers **add** the indices; to **divide** numbers **subtract** the indices; and to **raise a number to a power multiply** the indices. Hence $a^4 \times a^3 = a^7$ but $a^4 \times b^3 = a^4b^3$. Do not attempt to add the indices when the bases are different.

Compare (i) $2x$ (ii) x^2 (iii) $4x^2$ (iv) $(4x)^2$ when $x = 3$.

(i) $2x = 2\times 3 = 6$
(ii) $x^2 = x\times x = 3\times 3 = 9$
(iii) $4x^2 = 4\times x\times x = 4\times 3\times 3 = 36$
(iv) $(4x)^2 = 4x\times 4x = 12\times 12 = 144$

In particular notice how $4x^2$ differs from $(4x)^2$.

To simplify (i) $3x^2\times 2x^3$ and (ii) $16x^8\div 8x^2$
(i) can be written $3\times 2\times x^2\times x^3 = 6x^5$. Take care to **add** the indices and **multiply** the numbers.
(ii) can be written $\dfrac{16x^8}{8x^2} = 2x^6$. Take care to **subtract** the indices and **divide** the numbers.

Note that $(-2)^2 = -2\times -2 = +4$

$(-2)^3 = -2\times -2\times -2 = -8$

$(-2)^4 = -2\times -2\times -2\times -2 = +16$

and similarly $(-2)^5 = -32$.

A **negative** number taken to an **even** power is **positive** and taken to an **odd** power is **negative**.

When substituting numbers for letters, care is required with the signs. If $a = 3$, $b = -2$, $c = 0$ evaluate (i) a^2+b^2 (ii) a^2-b^2 (iii) $(4a)^2-3b^3$ (iv) $5a^2b^3c$ (v) $5a^2\div 3b$.
(i) $a^2+b^2 = (3)^2+(-2)^2 = 9+4 = 13$.
(ii) $a^2-b^2 = (3)^2-(-2)^2 = 9-4 = 5$.
(iii) $(4a)^2-3b^3 = (4\times 3)^2-3(-2)^3 = 12^2-3\times(-8)$
$= 144-(-24) = 144+24 = 168$.
(iv) $x\times 0 = 0$ for all values of x. $\therefore$ $5a^2b^3c = 0$, since $c = 0$.
(v) $5a^2\div 3b = (5\times 3\times 3)\div(3\times -2) = 45\div -6 = -7\frac{1}{2}$.

The **commutative law**: $a+b = b+a$ and $ab = ba$ are true for addition and multiplication, but $a-b \neq b-a$ and $a\div b \neq b\div a$.

The **associative law**: $(a+b)+c = a+(b+c)$ and $(ab)c = a(bc)$ are true but $(a\div b)\div c \neq a\div(b\div c)$ and $(a-b)+c \neq a-(b+c)$.

The **distributive law**: $a(b+c) = ab+ac$ is true.

$2x^2\times 5x^3$ can be simplified to $2\times 5\times x^2\times x^3 = 10x^5$, but $2x^2+5x^3$ cannot be simplified further because the x^2 and x^3 terms are **unlike**. The following simplifications are possible:

$2x^2 + 5x^2 = 7x^2$, the x^2 terms being **like** terms.

$3x^2 - 2x + 5x^2 - 4x = 8x^2 - 6x$ and no further, x^2 is not like to x.

$5x^2 \times -3xy = 5 \times -3 \times x^2 \times x \times y = -15x^3y$.

Brackets

From the **distributive law** $3(2x-4y) = 6x - 12y$;
$x(x+y) = x^2 + xy$.

An important case is $-(x-y)$. It implies $-1(x-y) = -1x-(-y)$ or $-x+y$ when the bracket is removed. $\therefore\ y-x = -(x-y)$.

This idea is used to simplify $(4x+5y)-(2x+3y) = 4x+5y-2x-3y$ (note the sign change) $= 2x+2y$.

To multiply two brackets $(a+b)(c+d)$: multiply each term of the first bracket by each term of the second.

$(a+b)(c+d) = ac + ad + bc + bd$.

$(3x+2y)(a-3b) = 3xa-9xb+2ya-6yb$. This can also be written $3ax+2ay-9bx-6by$, because $ab = ba$ and $(a+b)+c = a+(b+c)$.

$(x+2)(3x+4) = 3x \times x + 4x + 6x + 8 = 3x^2 + 10x + 8$.

Notice that there are two like terms and the expression reduces to three terms. An expression of the form ax^2+bx+c is called a **quadratic** expression.

There are two special cases of the quadratic:

(i) **the perfect square**

$$(a+b)^2 = (a+b)(a+b)$$
$$= a^2 + ab + ba + b^2 = a^2 + 2ab + b^2$$
$$(a-b)^2 = (a-b)(a-b)$$
$$= a^2 - ab - ba + b^2 = a^2 - 2ab + b^2$$

Notice that $ab = ba$ makes $+ab+ba = +2ab$, $-ab-ba = -2ab$.

(ii) **the difference between two squares**

$$(a+b)(a-b) = a^2 - ab + ba - b^2 = a^2 - b^2.$$

The ab and ba terms add up to zero, leaving a square minus another.

Example (i) Simplify $(x-4)^2-(3x+2)(x-5)$; (ii) if $x = 3$, $y = -2$ and $a = 4$: find the value of $(x-y)^2+a(2x+y)$.

(i) Rewrite as $(x-4)(x-4)-(3x+2)(x-5)$ and multiply out.

$$x^2 - 4x - 4x + 16 - (3x^2 - 15x + 2x - 10)$$
$$= x^2 - 8x + 16 - (3x^2 - 13x - 10)$$
$$= x^2 - 8x + 16 - 3x^2 + 13x + 10.$$

Notice that the second bracket is retained because of the $-$ sign outside. Change the signs accordingly when it is removed. Gather the like terms to give the answer $-2x^2+5x+26$.

(ii) When substituting numbers for letters, it is easier to substitute the numbers and work out the numerical value of each bracket first.

$$(3-(-2))^2 + 4(6+(-2)) = (3+2)^2 + 4(6+(-2))$$
$$= 5^2 + 4 \times 4 = 25 + 16 = 41.$$

Factors

In the same way that 3 and 4 are factors of 12, so 3 and $(x+y)$ are factors of $3x+3y$, because when multiplied together they give $3x+3y$. This is the **distributive law**; it is apparent here that 3 has been distributed.

To find the factors of $2\pi r^2+2\pi rh$ (i) look for any common distributed terms (here we have $2\pi r$); (ii) the factors will be of the form $2\pi r(\quad)$. To find the contents of the bracket, ask the question 'What must $2\pi r$ be multiplied by to give $2\pi r^2$?' The answer is r. Repeat the question for the other term to give the factors $2\pi r(r+h)$.

To factorise $-2x^2-6x$, notice that $-2x$ has been distributed as the common factor $\therefore$ $-2x^2-6x=-2x(x+3)$. Note the sign in the bracket.

To factorise $2a(x-y)+b(x-y)$, notice that the expression has a common bracket. $(x-y)$ is a factor. If this is removed $(2a+b)$ will be the other one. The factors are $(x-y)(2a+b)$.

To factorise $3ap-aq+3bp-bq$, factorise the first pair of terms separately from the second pair: $a(3p-q)+b(3p-q)$.
In this line there is a common bracket of $(3p-q)$.
Removing this makes the factors $(3p-q)(a+b)$.

Factorising quadratic expressions

From page 19 $(x-3)(x+4) = x^2 + 4x - 3x - 12 = x^2 + x - 12$

$$(x-6)(x-2) = x^2 - 2x - 6x + 12 = x^2 - 8x + 12$$

$\therefore$ the factors of $x^2 + x - 12 = (x-3)(x+4)$ (a)

and the factors of $x^2 - 8x + 12 = (x-6)(x-2)$. (b)

Study the contents of the brackets in each case:

(i) there are two brackets with an x in each;

(ii) in (a) the signs are different because the number term is -12; in (b) the signs are the same because the number term is $+12$;

(iii) in (a) the numbers used are 3 and 4 because $\mathbf{3 \times 4 = 12}$ and at the same time $\mathbf{4-3 = 1}$, which is the number of x; in (b) the numbers used are 6 and 2 because $\mathbf{6 \times 2 = 12}$ and at the same time $\mathbf{6+2 = 8}$, the number of x.

These points should help to reduce the amount of working, which is mainly trial and error.

To find the factors of $x^2+9x+14$

(i) write down the brackets $(x \quad)(x \quad)$;

(ii) the signs in each are the same (like signs multiplied give $+$);

(iii) for the numbers only $7 \times 2 = 14$ and $7+2 = 9$ at the same time.

$\therefore$ the factors could be:

$$(x-7)(x-2) = x^2 - 2x - 7x + 14 = x^2 - 9x + 14$$

$$(x+7)(x+2) = x^2 + 2x + 7x + 14 = x^2 + 9x + 14 \quad *$$

Multiplying each carefully it can be seen that only * gives the original expression $\therefore (x+7)(x+2)$ are the factors.

With $2x^2+x-15$ the number of possibilities is increased. The constant term is $-$ so the signs in the brackets will be different. This fact reduces the trials considerably. Try the following first:

$$(2x+5)(x-3) = 2x^2 - 6x + 5x - 15 = 2x^2 - x - 15$$

$$(2x-5)(x+3) = 2x^2 + 6x - 5x - 15 = 2x^2 + x - 15 \quad *$$

$$(2x+3)(x-5) = 2x^2 - 10x + 3x - 15 = 2x^2 - 7x - 15$$

$$(2x-3)(x+5) = 2x^2 + 10x - 3x - 15 = 2x^2 + 7x - 15$$

$\therefore$ from * the factors are $(2x-5)(x+3)$.

Had none of these produced the answer it would be necessary to try the combinations of $(2x \quad 15)(x \quad 1)$.

The factors of $x^2-14x+49 = (x-7)(x-7) = (x-7)^2$.
This is a perfect square. The feature to note is that $7^2 = 49$ and $7+7 = 14$. The constant term is ($\frac{1}{2}$ the x number)^2.

The factors of $4x^2-25 = (2x-5)(2x+5)$.
On page 19 this was referred to as the **difference between two squares**. The brackets contain the square root of each term, one with a plus sign, the other with a minus.

The factors of $3x^2 - 8x = x(3x-8)$.
This is a quadratic expression and the temptation to put down two brackets must be resisted. There is a common factor of x.

The factors of $x^2 - 3xy - 10y^2 = (x-5y)(x+2y)$.
When an expression has x^2 and y^2 terms put an x and a y in each bracket, then determine the numbers as before.

The factors of $5 - 6x - 8x^2 = (5+4x)(1-2x)$.
If the terms are in reverse order, there is no need to rearrange them. Put the x in the right hand side of each bracket and then proceed as before.

Fractions

In arithmetic $\frac{2}{3} \times \frac{5}{7} = \frac{10}{21}$, and in algebra $\frac{a}{b} \times \frac{c}{d} = \frac{ac}{bd}$. **Cancelling** is also carried out in a similar way.

$$\frac{a+ab}{a} = \frac{a(1+b)}{a} = 1 + b$$

To simplify this fraction, factorise the numerator as shown; a can be cancelled leaving $1+b$. In $\frac{a+b}{a}$ it is incorrect to cancel the a's. There is no common factor in the numerator and denominator. There is no simpler form.

In the case of $\frac{x^2-16}{x-4} = \frac{(x-4)(x+4)}{x-4} = x+4$, factorise the numerator first then cancel the $x-4$. (Do not cancel x^2 with x or 16 with 4).

To **add or subtract** algebraic fractions the common denominator must be found, as in arithmetic. If two fractions have denominators 21 and 35, factorise both. $21 = 7 \times 3$ and $35 = 7 \times 5$. The common denominator is $7 \times 3 \times 5 = 105$. It contains the 21 and the 35 as 7×3 and 7×5 respectively. If the denominators are $x^2 - 5x$ and $x^2 - 25$, by factorising $x^2 - 5x = x(x-5)$ and $x^2 - 25 = (x-5)(x+5)$ and the common denominator is $x(x-5)(x+5)$, each factor appearing once.

$$\frac{1}{a^3} + \frac{2}{a^2} - \frac{1}{a} = \frac{1}{a^3} + \frac{2a}{a^3} - \frac{a^2}{a^3} = \frac{1+2a-a^2}{a^3}$$

The common denominator is a^3. Express each denominator as a^3, then add the numerators.

$$\frac{1}{x^2-2x}-\frac{1}{x^2-3x+2}=\frac{1}{x(x-2)}-\frac{1}{(x-2)(x-1)}$$
$$=\frac{(x-1)}{x(x-2)(x-1)}-\frac{x}{x(x-2)(x-1)}$$
$$=\frac{x-1-x}{x(x-2)(x-1)}=\frac{-1}{x(x-2)(x-1)}$$

Factorise each denominator. The common one is $x(x-2)(x-1)$. Write each fraction with this denominator. Add the numerators.

Simple equations

$x+6=8$ is a **simple equation**. It is true for only one value of x. In this case $x=2$ and this is called the **solution** of the equation.

It will help at this stage to define the **inverse** elements.
The **additive inverse** of $+a$ is $-a$, because $+a-a=0$.
The **multiplicative inverse** of a is $\frac{1}{a}$ because $a\times\frac{1}{a}=1$.

These will be used in the following examples.

To solve $5x+3=4x+7$.
The x terms are placed on one side of the equation and the numbers on the other. To do this use the inverses mentioned above. Add -3 to both sides (the additive inverse of $+3$).

$\Leftrightarrow 5x+3-3=4x+7-3 \Leftrightarrow 5x=4x+4$

Now add $-4x$ to both sides (the additive inverse of $+4x$).

$5x-4x=4x+4-4x \Leftrightarrow x=4$ is the solution.

Adding or subtracting the same number from both sides of an equation balances it. This is also true if the equation is multiplied or divided on both sides by the same number.

To solve $\frac{2x}{3}=10 \Leftrightarrow \frac{3}{2}\times\frac{2}{3}x=\frac{3}{2}\times 10 \Leftrightarrow x=\frac{30}{2} \Leftrightarrow x=15$.
Multiply both sides by $\frac{3}{2}$, the inverse of $\frac{2}{3}$.

To solve $5(2x+3)-1=3(2x+7)$, first remove the brackets.

$\Leftrightarrow 10x+15-1=6x+21$.

$\Leftrightarrow 10x+14=6x+21$

$\Leftrightarrow 10x + 14 - 14 = 6x + 21 - 14$ subtracting 14 from both sides.

$\Leftrightarrow 10x = 6x + 7$

$\Leftrightarrow 10x - 6x = 6x + 7 - 6x$ subtracting $6x$ from both sides.

$\Leftrightarrow 4x = 7 \Leftrightarrow x = \frac{7}{4}$ or $1\frac{3}{4}$ dividing both sides by 4.

To solve $\frac{1}{2}(2x+5)-\frac{1}{3}(x+3) = 4$, multiply both sides by the common denominator $2 \times 3 = 6$, in order to clear the fractions.

$6(\frac{1}{2}(2x+5)-\frac{1}{3}(x+3)) = 6 \times 4$

$\Leftrightarrow \frac{6}{2}(2x+5) - \frac{6}{3}(x+3) = 24$

$\Leftrightarrow 3(2x+5) - 2(x+3) = 24$ (the fractions are now eliminated)

$\Leftrightarrow 6x + 15 - 2x - 6 = 24$ multiplying out the brackets.

$\Leftrightarrow 4x + 9 = 24$ gathering like terms.

$\Leftrightarrow 4x + 9 - 9 = 24 - 9$ subtracting 9 from both sides.

$\Leftrightarrow 4x = 15 \Leftrightarrow x = \frac{15}{4}$ or $3\frac{3}{4}$ dividing both sides by 4.

Simultaneous equations

To solve $4x+3y = 5$, $3x-2y = 8$ it is necessary to know the values of x and y which satisfy the equations at the same time, **simultaneously**. Number the equations (i) and (ii). Multiply (i) by 2 and (ii) by 3, making the y terms the same in both equations.

$4x + 3y = 5$ (i)
$3x - 2y = 8$ (ii)

$$\begin{aligned} 8x + 6y &= 10 \\ 9x - 6y &= 24 \\ \hline 17x \qquad &= 34 \Leftrightarrow x = 2 \end{aligned}$$

Adding the equations eliminates the y terms. To find the value of y, put $x = 2$ in (i) or (ii) according to which is 'easier'. In (i)

$4 \times 2 + 3y = 5 \Leftrightarrow 8 + 3y = 5 \Leftrightarrow 3y = 5 - 8 \Leftrightarrow 3y = -3$

$\Leftrightarrow y = -1$

$\therefore$ the solution is $x = 2$, $y = -1$.

Note that (i) and (ii) are equations of **straight lines**, which in general meet in one point. In this case it is $(2, -1)$. A single equation with two unknowns, e.g. $3x-2y = 8$, has an infinite number of solutions, representing all the points on the line $(0, -4)$ $(2, -1)$ $(4, 2)$, etc.

Two parallel lines $3x-2y = 5$ and $3x-2y = 8$ do not meet and will have no solutions.

To solve $9x+5y=-13$, $6x+3y=-9$ multiply (i) by 2 and (ii) by 3.

$$\begin{aligned} 9x + 5y &= -13 \quad \text{(i)} \qquad & 18x + 10y &= -26 \\ 6x + 3y &= -\ 9 \quad \text{(ii)} \qquad & 18x + \ 9y &= -27 \end{aligned}$$

subtract $\quad y = -26 + 27 \Leftrightarrow y = 1.$

Substitute this in (ii)

$6x + 3 = -9 \Leftrightarrow 6x = -9 - 3 \Leftrightarrow 6x = -12 \Leftrightarrow x = -2$

the solution is $(-2, 1)$.

In the first example the equations are **added** because the $6y$ terms have **different** signs. In the second they are **subtracted** because the $18x$ terms have the **same** sign.

Quadratic equations

An equation of the form $2x^2-5x-5=0$ which contains a quadratic expression is called a **quadratic equation**. The method of solution is based on the assumption that if two quantities a and b are such that $\boldsymbol{ab = 0}$ then either $\boldsymbol{a = 0}$ **or** $\boldsymbol{b = 0}$.

To solve $x^2-3x-4=0$, factorise the left-hand side:

$(x-4)(x+1) = 0$

The product of the factors is 0, $\Rightarrow$ either $x-4=0$ or $x+1=0$.

If $x - 4 = 0$ then $x = 4$ and if $x + 1 = 0$ then $x = -1$

$x = 4$ and -1 are called the **solutions** or **roots** of the equation. When substituted in the equation they make the left-hand side 0.

To solve $3x^2-2x=8 \Leftrightarrow 3x^2-2x-8=0$

$\Leftrightarrow (3x+4)(x-2) = 0 \Leftrightarrow$ either $3x + 4 = 0$ or $x - 2 = 0$.

$3x + 4 = 0 \Leftrightarrow 3x = -4 \quad \boldsymbol{x = -\tfrac{4}{3}}. \quad x - 2 = 0 \Leftrightarrow \boldsymbol{x = 2}$

The right-hand side of the equation must be 0.

To solve $3x^2 = 18x. \Leftrightarrow 3x^2-18x=0 \Leftrightarrow 3x(x-6)=0$

$\Leftrightarrow 3x = 0$ or $x - 6 = 0 \Leftrightarrow \boldsymbol{x = 0}$ or $\boldsymbol{x = 6}$.

Do not cancel the x's initially. The root $x=0$ will be lost. Make the right-hand side 0, then take out the common factor.

To solve $x^2=49 \Leftrightarrow x^2-49=0 \Leftrightarrow (x-7)(x+7)=0$

$\Leftrightarrow x - 7 = 0$ or $x + 7 = 0 \Leftrightarrow \boldsymbol{x = +7}$ or $\boldsymbol{-7}$ written $\boldsymbol{\pm 7}$

Or, square-root both sides to give $x = \pm 7$, remembering that $(-7)^2 = 49$ so both $+$ and $-$ 7 are needed.

To solve $\dfrac{x-2}{x} = \dfrac{x-3}{2} \qquad \Leftrightarrow 2(x-2) = x(x-3)$

$\Leftrightarrow 2x - 4 = x^2 - 3x \Leftrightarrow x^2 - 5x + 4 = 0 \Leftrightarrow (x-4)(x-1) = 0$

$\Leftrightarrow x - 4 = 0$ or $x - 1 = 0 \Leftrightarrow$ $\mathbf{x = 4}$ **or 1**

Multiply both sides by $2x$ to eliminate the fractions. Removing the brackets reveals a quadratic equation. Rearrange and solve as usual.

Solution by formula

To solve $2x^2+3x-6 = 0$ to two decimal places.

This expression does not factorise so use the formula related to

$$ax^2 + bx + c = 0 \qquad x = \frac{-b \pm \sqrt{b^2-4ac}}{2a}$$

In our equation $a = 2, b = 3$ and $c = -6$. Substitute in the formula:

$$x = \frac{-(3) \pm \sqrt{(3)^2 - 4 \times 2 \times (-6)}}{2 \times 2} = \frac{-3 \pm \sqrt{9+48}}{4} = \frac{-3 \pm \sqrt{57}}{4}$$

$$x = \frac{-3 \pm 7{\cdot}550}{4} \text{(using the root table)} \quad x = \frac{4{\cdot}55}{4} \quad \text{or} \quad \frac{-10{\cdot}55}{4}$$

$x = 1{\cdot}137\,5$ or $-2{\cdot}637\,5$ or $1{\cdot}14$ or $-2{\cdot}64$ to two decimal places.

(Check with the syllabus that this topic is included. It only appears on a few. This formula can be used to solve all quadratics, but to solve 'to 2 decimal places' suggests that it must be used.)

Rearrangement of formulae

If $A = p+q$ then $A-q = p$. p has become the subject by subtracting q from both sides (the inverse of $+q$). If $A = pq$ then $\frac{A}{q} = p$. p has become the subject by dividing through by q.

The same processes as those used in solving equations, including squaring and square-rooting both sides, are used.

To make r the subject of $x = pq+5r \Leftrightarrow x-pq = 5r \Leftrightarrow \frac{1}{5}(x-pq) = r$ or $r = \frac{1}{5}(x-pq)$.

Subtract pq from both sides, $\div$ through by 5 and r is the subject.

To make x the subject of

$$A = \frac{x}{x-1}. \Leftrightarrow A(x-1) = \frac{x(x-1)}{x-1} \Leftrightarrow A(x-1) = x \Leftrightarrow Ax - A = x.$$

$$\Leftrightarrow Ax - x - A = 0 \Leftrightarrow Ax - x = A \Leftrightarrow x(A-1) = A$$

$$\Leftrightarrow x = \frac{A}{A-1}$$

Multiply both sides by $(x-1)$ and remove the brackets. x now appears on both sides of the formula. Rearrange with both x terms on the same side. Take out the common factor of x, then divide both sides by $(A-1)$. x is now the subject.

To make x the subject of

$$A = 2\sqrt{(x^2+4)} \Leftrightarrow A^2 = 4(x^2+4) \Leftrightarrow A^2 = 4x^2 + 16$$

$$\Leftrightarrow A^2 - 16 = 4x^2 \Leftrightarrow \frac{A^2-16}{4} = x^2$$

$$\Leftrightarrow x = \pm\sqrt{\frac{A^2-16}{4}}$$

Square both sides (remember the 2), subtract 16 from both sides, divide by 4, and finally square-root both sides.

$$\frac{a}{b} = \frac{c}{d} \Leftrightarrow ad = bc \Leftrightarrow d = \frac{bc}{a} \Leftrightarrow \frac{d}{b} = \frac{c}{a} \Leftrightarrow \frac{1}{bc} = \frac{1}{ad}$$

When two ratios are given equal the top-left term can be cross-multiplied with the bottom-right term and the top-right with the bottom-left. All of the above arrangements are a result of cross-multiplying. Do not apply it to $\frac{1}{a}+\frac{1}{b}=\frac{1}{c}$. Only one fraction is allowed on each side of the equals sign.

Key terms

An **index** is a power.
Like terms have the same letters and powers. E.g. $3x^2y$ and $15x^2y$.
Unlike terms do not have the same letters and powers. E.g. $3xy^2$ and $15x^2y$ are unlike terms.
The **perfect square** is of the form $(a+b)^2$ and $(a-b)^2$.
The **difference between two squares** is of the form a^2-b^2.
The **sum of two squares** is of the form a^2+b^2.
The **additive inverse** of $+a$ is $-a$ and *vice versa* such that $+a+(-a)=0$.
The **multiplicative inverse** of a is $\frac{1}{a}$ such that $a \times \frac{1}{a} = 1$.
A **quadratic** expression is of the form ax^2+bx+c.
The **roots** or **solutions** of an equation are the numbers which satisfy the equation.

Chapter 3
Number Systems

Our number system is called the Hindu-Arabic system. The symbols used are called **digits**: 0, 1, 2, 3, 4, 5, 6, 7, 8, 9. The numbers they form can be listed in the following sets:

Integers $\{\ldots -3, -2, -1, 0, +1, +2, +3 \ldots\}$
There are negative and positive integers, but 0 is neither positive nor negative. It is just an integer.

Natural numbers $\{1, 2, 3, 4, 5, 6 \ldots\}$

Whole numbers $\{0, 1, 2, 3, 4, 5 \ldots\}$
0 is a whole number, but is not considered to be natural. Both whole and natural numbers are subsets of the set of integers.

Rational numbers $\{\frac{1}{3}, \frac{1}{2}, \frac{3}{4}, 0{\cdot}7, \frac{0}{4}, \frac{3}{1}, \frac{17}{2} \ldots\}$
are numbers which can be expressed as a fraction or a **ratio**. $0{\cdot}7 = \frac{7}{10}$; $\frac{3}{1} = 3$; $\frac{0}{4} = 0$.

Irrational numbers $\{\pi, \sqrt{2}, \sqrt{3}, \sqrt{5}, \sqrt{6} \ldots\}$
are numbers which cannot be expressed as fractions or ratios. For over one thousand years mathematicians have tried to find the exact value of π. Computers have found the value to thousands of decimal places, but the decimal has never ended, nor has it started to recur. π is known to lie between 3·141 and 3·142, to 7 decimal places 3·141 592 6, so the value $\frac{22}{7}$, which is 3·142 857 1 to 7 decimal places, is not the best approximation. Similarly $\sqrt{2}, \sqrt{3}$, etc. cannot be expressed exactly. In contrast $\frac{1}{3}$, although a decimal which does not end, is known to be 0·3 recurring. It is a known fraction $\therefore$ is rational.

Prime numbers $\{2, 3, 5, 7, 11, 13 \ldots\}$
are numbers which are divisible by 1 and themselves only. 2 is the first prime, 1 is excluded from the set.

Real numbers are all the numbers which can be placed on the number line, i.e. integers, natural numbers, whole numbers, rational numbers and irrational numbers.

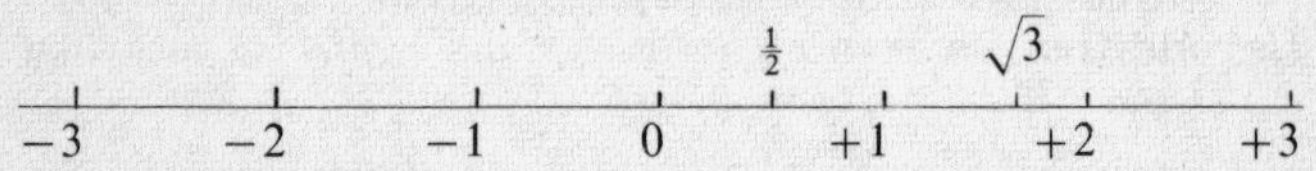

Example Given that n is a positive integer, prove that the integer $n^2 + n$ is always even.

It is not sufficient to verify this for a few values chosen at random. Consider $n^2 + n$ where n is a positive integer.

It can be written $n(n+1)$.
If n is **odd**, then $(n+1)$ is **even** and the product of an odd number and an even number is even.
If n is **even**, then $(n+1)$ is **odd** and the product of an even number and an odd number is even.
$\therefore n^2+n$ is even for all positive integral values of n.

Example Find the values of x and y which belong to the set of natural numbers satisfying $3x+4y \leqslant 13$.

Systematically substitute values of x and y into the inequality starting with $x = 1$ and $y = 1$, 0 not being a natural number.

$x = 1$	$y = 1$	$3 + 4 \leqslant 13$	is true
$x = 1$	$y = 2$	$3 + 8 \leqslant 13$	is true
$x = 1$	$y = 3$	$3 + 12 \leqslant 13$	is false
$x = 2$	$y = 1$	$6 + 4 \leqslant 13$	is true
$x = 2$	$y = 2$	$6 + 8 \leqslant 13$	is false
$x = 3$	$y = 1$	$9 + 4 \leqslant 13$	is true
$x = 3$	$y = 2$	$9 + 8 \leqslant 13$	is false

There are no more natural number values. From this point they are all too large. (1, 1), (1, 2), (2, 1), (3, 1) is the set of solutions written as co-ordinates.

Different number bases
Our ordinary numbers are called decimal or **denary** numbers. The column headings are the powers of 10:

... 10000	1000	100	10	1
10^4	10^3	10^2	10^1	10^0

With **binary** numbers the headings are powers of 2 (Base 2).

... 128	64	32	16	8	4	2	1
2^7	2^6	2^5	2^4	2^3	2^2	2^1	2^0

Base 3 numbers have the powers of 3 as column headings:

... 243   81   27   9   3   1

Octal (base 8) numbers, like binary numbers, appear more often, owing to their use in computers. The headings are ... 512 64 8 1

The list overleaf shows some easy numbers in different bases.

Denary	Binary					Base 3			Octal		
	16	8	4	2	1	9	3	1	64	8	1
1					1			1			1
2				1	0			2			2
3				1	1		1	0			3
4			1	0	0		1	1			4
5			1	0	1		1	2			5
6			1	1	0		2	0			6
7			1	1	1		2	1			7
8		1	0	0	0		2	2		1	0
15		1	1	1	1	1	2	0		1	7
23	1	0	1	1	1	2	1	2		2	7

To **convert** the **denary** number 115 to **binary**, use either of the following methods:

(i)

64	32	16	8	4	2	1
1	1	1	0	0	1	1

115
64 Put a 1 in the 64 column
51
32 Put a 1 in the 32 column
19
16 Put a 1 in the 16 column
3
2 Put a 1 in the 2 column
1 Put a 1 in the 1 column

(ii)

2	115
2	57 r 1
2	28 r 1
2	14 r 0
2	7 r 0
2	3 r 1
2	1 r 1
2	0 r 1

$115_{10} = 1\,110011_2$

(i) Write down the headings, build up the number 115 by placing 0 or 1 in the correct columns. Keep a running total as shown.
(ii) Divide 115 successively by 2, recording the remainders. From the bottom to the top gives the number from left to right.

To convert $1\,101\,101_2$ to a denary number, write this number down under binary headings. Find the sum of the headings with the 1's.

64	32	16	8	4	2	1	
1	1	0	1	1	0	1	$64 + 32 + 8 + 4 + 1 = 109$.

Binary addition and subtraction

The following three binary additions are the basis for all binary additions. $1+0 = 1$, $1+1 = 10$ which means 'put down 0 and carry 1 to the next column; and $1+1+1 = 11$ which means 'put down 1 and carry 1 to the next column'.

To evaluate (i) $1\,111_2 + 1\,011_2$; (ii) $10\,111_2 + 1\,111_2 + 111_2$:

$$\text{(i)}\quad \begin{array}{r} 1\,111 \\ 1\,011 \\ \hline 11\,010 \end{array} \qquad \text{(ii)}\quad \begin{array}{r} 10\,111 \\ 1\,111 \\ \hline 100\,110 \\ 111 \\ \hline 101\,101 \\ \hline \end{array}$$

(i) In the unit column $1+1 = 10$; put down 0 and carry 1. In the 2's column $1+1+1 = 11$; put down 1 and carry 1. In the 4's column $1+0+1 = 10$; put down 0 and carry 1. In the 8's column $1+1+1 = 11$.

(ii) When adding three or more numbers, add the first two and then add on the third, etc. This can save difficulty.

Subtraction is based on $1-0 = 1 : 1-1 = 0 : 10-1 = 1$. When 'borrowing' 1 remember that it becomes 2 in the next column. It is easier to make the adjustment to the top line when borrowing.

To evaluate (i) $11\,101 - 1\,110$; (ii) $10\,000 - 111$:

$$\text{(i)}\quad \begin{array}{r} 11\,101 \\ -\ 1\,110 \\ \hline 1\,111 \\ \hline \end{array} \qquad \text{(ii)}\quad \begin{array}{r} 1\ 1\ 1\ 10 \\ 1\ 0\ 0\ 0\ 0 \\ -\quad 1\ 1\ 1 \\ \hline 1\ 0\ 0\ 1 \\ \hline \end{array}$$

(i) In the units column $1-0 = 1$. In the 2's column $0-1$ is not possible, so we borrow 1 from the 4's column. Then $10-1 = 1$. The top digit in the 4's column is now 0; borrow 1 again from the 8's column: then $10-1 = 1$ etc.

(ii) The case where the top line contains a number of 0's. Here the 16 is written $8+4+2+2$, which in binary is written above the 10 000. Subtract 111 from that line.

Binary multiplication and division

The binary multiplication table is $1 \times 0 = 0$: $0 \times 1 = 0$: $0 \times 0 = 0$: $1 \times 1 = 1$.

Multiplying by 2 in binary has the same effect as multiplying by ten in denary. In binary $111 \times 10 = 1\,110$, similarly $111 \times 100 = 11\,100$.

To evaluate $1\,101 \times 111$ multiply in the usual way. There are three numbers to add. Proceed as shown before.

```
   1 101          The addition      1 101
     111                           11 010
  --------                        --------
   1 101                          100 111
  11 010                          110 100
 110 100                          --------
 --------                         1011011
 1011011
```

To evaluate 100 010 ÷ 101, treat as long division. If 101 goes into a binary number it goes 1; if it does not go write 0.

```
       110 remainder 100
     ________________
101 | 100010
       101
       ---
       111
       101
       ---
        100
```

100 010 = 34; 101 = 5
34 ÷ 5 = 6 remainder 4
110 = 6; 100 = 4.
The binary calculation is correct.

Although the working is supposed to be shown in binary, a check can be made in denary as shown.

Dividing by 2 in binary has the same effect as dividing by ten in denary. In binary 1 110 ÷ 10 = 111 and similarly 11 000 ÷ 1 000 = 11.

Bicimals are the fractional parts of binary numbers. The headings appear thus:

decimal	bicimal
$\ldots 100\ 10\ 1 \cdot \frac{1}{10}\ \frac{1}{100}\ \frac{1}{1000}\ \ldots$	$\ldots 8\ 4\ 2\ 1 \cdot \frac{1}{2}\ \frac{1}{4}\ \frac{1}{8}\ \frac{1}{16}\ \ldots$

To convert $101{\cdot}101\,1_2$ to a decimal, place the digits under the correct headings and add.

$4\ 2\ 1 \cdot \frac{1}{2}\ \frac{1}{4}\ \frac{1}{8}\ \frac{1}{16} \qquad 4 + 1 \cdot \frac{1}{2} + \frac{1}{8} + \frac{1}{16}$

$1\ 0\ 1 \cdot 1\ 0\ 1\ 1 \qquad 5 + 0{\cdot}5 + 0{\cdot}125 + 0{\cdot}0625 = 5{\cdot}6875$

To convert (i) $7\frac{5}{8}$ (ii) $7\frac{3}{5}$ to bicimals. (i) is straightforward because $\frac{5}{8}$ can be split into $\frac{1}{2}$, $\frac{1}{4}$ etc., but not so with (ii), where we change the 3 and 5 to binary and perform binary division.

(i) $7\frac{5}{8} = 7 + \frac{1}{2} + \frac{1}{8}$

$= 111{\cdot}101$

(ii) $\frac{3}{5} = \frac{11}{101}$

```
       0·100110
     __________
101 | 11·000000
      10 1
      ----
        1000
         101
        ----
          110
          101
          ---
           10
```

$\therefore 7\frac{3}{5} = 111{\cdot}10011$ to 5 bicimal places.

Working in other bases

To convert 174 (i) to base 3 and (ii) to base 5, carry out successive division by 3 and 5 respectively; the remainders will supply the answers in each case. (Bottom to top reads left to right.)

(i)

3	174	
3	58	*r* 0
3	19	*r* 1
3	6	*r* 1
3	2	*r* 0
3	0	*r* 2

(ii)

5	174	
5	34	*r* 4
5	6	*r* 4
5	1	*r* 1
5	0	*r* 1

$174_{10} = 20\,110_3$ and 1144_5

To convert $1\,247_8$ to denary, write down the headings as powers of 8, place 1, 2, 4 and 7 in the correct columns. Add as shown.

512	64	8	1
1	2	4	7

$$1\ 2\ 4\ 7 = (512\times 1)+(64\times 2)+(8\times 4)+(1\times 7)$$
$$= 512+128+32+7 = 679$$

$\therefore 1\,247_8 = 679$ in denary form.

To evaluate (i) 341_6+245_6; (ii) 321_4-123_4:

(i)
$$\begin{array}{r} 341 \\ +\ 245 \\ \hline 1\,030 \\ \hline \end{array}$$

(ii)
$$\begin{array}{r} 321 \\ -123 \\ \hline 132 \\ \hline \end{array}$$

(i) In base 6, $5+1=10$; put down 0 carry 1. In the next column $4+4+1=9$; put down 3 and carry 1. Finally $3+2+1=10$.
(ii) $1-3$ requires a borrowing of 4 from the next column. Then $5-3=2$. Proceed in this way remembering to borrow 4 each time.

To convert 237_8 to binary can be carried out without converting to base ten first. $2^3 = 8$, so for every octal heading there are three binary headings.

Octal		64			8			1
		2			3			7
Binary	128	64	32	16	8	4	2	1
	1	0	0	1	1	1	1	1

The 7 units produce the binary number 111, 3 in the 8 column is 011 and 2 in the 64 column is the binary number 10.
Hence $237_8 = 10\,011\,111_2$.

Example (i) The following sums are correct. In which bases are they worked?

(a)
$$\begin{array}{r} 245 \\ +134 \\ \hline 412 \\ \hline \end{array}$$

(b)
$$\begin{array}{r} 231 \\ -142 \\ \hline 45 \\ \hline \end{array}$$

(ii) Give an expression in terms of n for the value of the number 134_n in base 10. Find the value of n for which $134_n = 32_{10}$.

(i) (a) In denary $5+4 = 9$. 2 has been written down so 7 must have been carried. The sum is in base 7. This can be verified by pursuing the addition further.

(b) $1-2$ suggests that there has been some borrowing. In denary $7-2 = 5$ or $(6+1)-2 = 5$. 6 has been borrowed. The sum is in base 6. This can be confirmed by continuing the working.

(ii) Put down the column headings

n^2	n	1
1	3	4

in base 10 the number is n^2+3n+4. We must solve the quadratic equation $n^2+3n+4 = 32$ by factorising (see page 20).

$n^2+3n+4 = 32 \Leftrightarrow n^2+3n-28 = 0 \Leftrightarrow (n+7)(n-4) = 0$
$\Leftrightarrow n+7 = 0$ or $n-4 = 0 \Leftrightarrow n = -7$ or 4 $\therefore$ $\boldsymbol{n = 4}$ only, because a number base is positive.

Modulo arithmetic

This is sometimes called **clock arithmetic**. In the case of **modulo** 5 it is assumed that the only numbers are the digits $\{0, 1, 2, 3, 4\}$. For purposes of calculating in modulo 5 place these digits on a clock face as in figure 7.

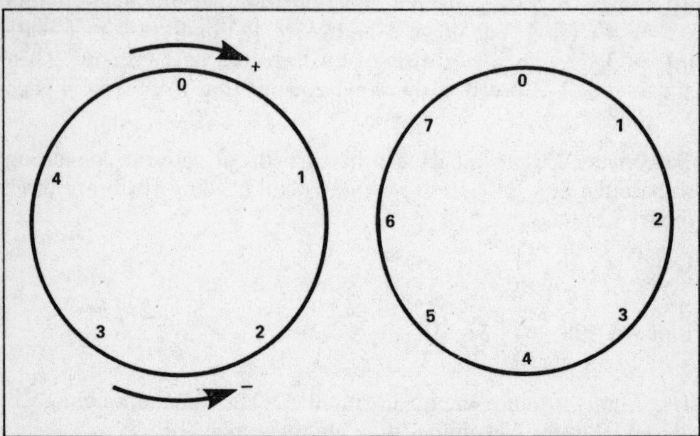

Figure 7

Addition and subtraction

To evaluate $4+3$ in modulo 5. Assume first clock in figure 7 has a hand pointing at 0. Describe 4 'hours' clockwise. Follow this with a further 3 'hours' clockwise. The hand will be pointing at 2, $\therefore 4+3 = 2$ in modulo 5. Verify that $3+3 = 1$ and $4+4 = 3$.

The commutative law holds for addition, i.e. $4+3 = 3+4 = 2$. The associative law holds for addition: $(1+3)+2 = 4+2 = 1$ and $1+(3+2) = 1+0 = 1 \therefore (1+3)+2 = 1+(3+2)$.

To evaluate $2-4$, point the hand at 0, move clockwise 2 and follow this by a movement of 4 anti-clockwise. The hand now points at 3. $\therefore 2-4 = 3$ in modulo 5.

Multiplication

To evaluate 3×2 in modulo 5, point the hand at 0 and move through 3 units clockwise, twice. The hand now points at 1, $\therefore 3 \times 2 = 1$. Verify that $2 \times 4 = 3$, $4 \times 4 = 1$ and $3 \times 3 = 4$.

The commutative law holds for multiplication, $2 \times 3 = 3 \times 2 = 1$. The associative law also holds.
Verify that $(2 \times 3) \times 4 = 2 \times (3 \times 4) = 4$.

Example In modulo 8 arithmetic work out the values of (i) $2+4+5$; (ii) $3-6$; (iii) 6×4; (iv) -5×3.

Draw the second clock in figure 7. Use the digits 0, 1, 2, 3, 4, 5, 6, 7.

(i) Move the hand through 2 followed by 4 and then 5 clockwise. The hand will point at 3, $\therefore 2+4+5 = 3$.
(ii) Move 3 clockwise followed by 6 anti-clockwise: $3-6 = 5$.
(iii) Describe 6 units 4 times. The hand points at 0. $6 \times 4 = 0$.
(iv) Describe 5 units anti-clockwise 3 times. The result is 1, $\therefore -5 \times 3 = 1$.

Example Form the multiplication table for modulo 6. Use it to solve the following equations: (i) $5x = 2$; (ii) $2x = 4$; (iii) $x^2 = 1$; (iv) $x^2 = x$, where x is a member of $\{0, 1, 2, 3, 4, 5\}$.

×	0	1	2	3	4	5
0	0	0	0	0	0	0
1	0	1	2	3	4	5
2	0	2	4	0	2	4
3	0	3	0	3	0	3
4	0	4	2	0	4	2
5	0	5	4	3	2	1

Draw a clock with the numbers 0, 1, 2, 3, 4, 5. Multiply them in pairs writing the results in the table. All the elements in the row and column headed 0 are 0. $2 \times 4 = 4 \times 2 = 2$ so the result at the intersection of row 2 column 4 is 2; similarly for row 4 column 2.

(i) For $5x = 2$ consult line 5 of the table. $5 \times 4 = 2$, $\therefore x = 4$.

(ii) For $2x = 4$ consult line 2. The number 4 appears under columns 2 and 5. $2 \times 2 = 4$ and $2 \times 5 = 4$, $\therefore x = 2$ and 5 are the two solutions.

(iii) For $x^2 = 1$ find a number squared with the result 1. $1 \times 1 = 1$ and $5 \times 5 = 1$, $\therefore x = 1$ and 5.

(iv) For $x^2 = x$, the square of the number equals the number. $1 \times 1 = 1$, $3 \times 3 = 3$ and $4 \times 4 = 4$, $\therefore x = 1, 3$ and 4.

Binary operations

The processes of addition, subtraction, multiplication, division, etc. are called operations. When such an operation is carried out on two numbers from a given set, it is called a **binary operation.**

Closure

Consider the set of integers $I = \{\ldots -3, -2, -1, 0, 1, 2, 3 \ldots\}$ and let the operation be addition. When any two members of I are added the result is also a member of I. E.g. $-3+2 = -1, 5+7 = 12$. The set I is said to be **closed** under the operation addition. It can be seen that I is also closed under multiplication $(3 \times -2 = -6)$. However under division I is not closed. E.g. $2 \div 5 = 0{\cdot}4$, the result not belonging to I.

Identity element

The **identity element** of a set is the number which, when combined with any member of the set under a given operation leaves the value of the member unaltered.

The identity element for addition is 0 because $4+0 = 4$; the identity element for multiplication is 1 because $4 \times 1 = 4$. The set I above has 0 as its identity element for addition and 1 for multiplication.

Not every set has its own identity element. E.g. The natural numbers $\{1, 2, 3, 4 \ldots\}$ do not possess the element 0, so the set does not possess the identity element for addition.

Inverse element

The **inverse** of a number under a given operation is the number which combined with a member produces the identity element of the operation. For addition -3 is the inverse of $+3$, because $(3)+(-3)=0$ and 0 is the identity element of addition. In multiplication $\frac{1}{3}$ is the inverse of 3 because $\frac{1}{3}\times 3=1$, the identity element of multiplication.

In the case of the set I, the integers, when the operation is addition, every element has an inverse which is also a member of I. $(-3)+(+3)=0$; $(+6)+(-6)=0$ etc., but when the operation is multiplication no inverse is an integer. $3\times\frac{1}{3}=1$; $4\times\frac{1}{4}=1$ etc.

The associative law

The symbol used to denote any operation in general is $*$. The **associative law** for an operation $*$ is: $(a*b)*c=a*(b*c)$.

Group

A set, S, which under an operation is (1) **closed**, (2) has an **identity** element which is a member of S, (3) for each element there is an **inverse** which is also an element of S, (4) is **associative**, is called a **group.**

If in addition to this the operation is **commutative**, i.e. $a*b=b*a$, it is called a commutative or Abelian group.

Questions are set to test the candidate's ability to apply an unusual operation.

Example If $a*b$ denotes the operation $2a-b$ evaluate (i) $5*3$; (ii) $(7*4)*2$ (iii) $7*(4*2)$ (iv) $y*y$ (v) find x if $8*x=10$.

(i) With $a=5$ and $b=3$; $5*3=(2\times 5)-3=10-3=7$.
(ii) The bracket first gives $(7*4)=(2\times 7)-4=14-4=10$.
$(7*4)*2=10*2=(2\times 10)-2=20-2=18$.
(iii) The bracket $(4*2)=(2\times 4)-2=8-2=6$.
$7*(4*2)=7*6=(2\times 7)-6=14-6=8$.
Notice that $(7*4)*2\neq 7*(4*2)$ so the operation is not associative.
(iv) $y*y=2y-y=y$
(v) $8*x=10\Leftrightarrow(2\times 8)-x=10\Leftrightarrow 16-x=10$
$\Leftrightarrow x=16-10\Leftrightarrow x=6$.

Questions on binary operations vary from those merely requiring the candidate to point out an inverse element, to those asking him to show that a set forms a group. The following question covers all the possibilities.

Example $a * b$ is defined as the remainder when $a+b$ is divided by 5. If $S = \{0, 1, 2, 3, 4\}$: (i) Construct a table to show the operation $*$ for S; (ii) give an example to show that $*$ is associative; (iii) assuming the associative law, show that the set forms a group under $*$.

(i) Here are some of the results of the operation $*$.

$0 * 1 = (0+1) \div 5 = 0$ remainder 1. Put 1 in line 0 column 1.
$1 * 3 = (1+3) \div 5 = 0$ remainder 4. Put 4 in line 1 column 3.
$1 * 4 = (1+4) \div 5 = 1$ remainder 0. Put 0 in line 1 column 4.
$4 * 4 = (4+4) \div 5 = 1$ remainder 3. Put 3 in line 4 column 4.

Continue the calculations for all pairs of S to complete the table.

$*$	0	1	2	3	4
0	0	1	2	3	4
1	1	2	3	4	0
2	2	3	4	0	1
3	3	4	0	1	2
4	4	0	1	2	3

(ii) To show that $(a * b) * c = a * (b * c)$ for three elements of S. Try $(2 * 3) * 4$; $(2 * 3) = 0$ and $0 * 4 = 4$ $\therefore$ $(2 * 3) * 4 = 4$.
Now consider $2 * (3 * 4)$. $(3 * 4) = 2$ and $2 * 2 = 4$ $\therefore$ $2 * (3 * 4) = 4$. $2 * (3 * 4) = (2 * 3) * 4 = 4$ is an example of $*$ being associative.

(iii) All members of the table belong to S so the set S is closed. Is there an identity element? Line 0 of the table tells us that $0 * 0 = 0; 0 * 1 = 1; 0 * 2 = 2; 0 * 3 = 3; 0 * 4 = 4$. The value of each element is unaltered when combined with 0. 0 is the identity element of S.
Does each element have an inverse element, which is also an element of S?
$0 * 0 = 0$. 0 is the identity element of $*$ $\therefore$ 0 is the inverse of 0.
$1 * 4 = 0$. 0 is the identity element of $*$ $\therefore$ 4 is the inverse of 1.
$2 * 3 = 0$ $\therefore$ 3 is the inverse of 2.
$3 * 2 = 0$ $\therefore$ 2 is the inverse of 3.
$4 * 1 = 0$ $\therefore$ 1 is the inverse of 4.
Each element has an inverse which is an element of S.
To sum up: the set is closed, it has an identity element, each element has an inverse in S, we assume the associative law, so the set forms a group under the operation $*$.

Isomorphic groups

Two groups are **isomorphic** if they have the same structure or form. When the tables are drawn up, corresponding results appear.

Example The set $S = \{a, b, c, d\}$ under the operation $*$ produces the table (A) as shown. Form the table for addition in modulo 4. (i) Show that modulo 4 addition forms a group; (ii) show that S forms a group under $*$ assuming the associative law; (iii) show that the two groups are isomorphic.

(A)

$*$	a	b	c	d
a	a	b	c	d
b	b	c	d	a
c	c	d	a	b
d	d	a	b	c

(B)

+	0	1	2	3
0	0	1	2	3
1	1	2	3	0
2	2	3	0	1
3	3	0	1	2

To draw up the table for addition in modulo 4, draw the clock face with the digits $\{0, 1, 2, 3\}$. Notice that $2+2 = 0$; $2+3 = 1$ etc. Table (B) shows the operation completed.

(i) Addition in modulo 4 is closed. Each element of the table is a member of the set $\{0, 1, 2, 3\}$.
The associative law holds for modulo 4 addition.
$0+0 = 0$; $0+1 = 1$; $0+2 = 2$; $0+2 = 3$; $\therefore$ 0 is the identity element.
$0+0 = 0$; $1+3 = 0$; $2+2 = 0$; $3+1 = 0$, where 0 is the identity element $\therefore$ each element has an inverse which is also a member of $0, 1, 2, 3$; $\therefore$ modulo 4 forms a group under addition.

(ii) Table (A) shows that S is closed under $*$. All the elements belong to the set $\{a, b, c, d\}$.
$a*a = a$; $a*b = b$; $a*c = c$; $a*d = d$. $\therefore$ a is the identity element.
$a*a = a$; $b*d = a$; $c*c = a$; $d*b = a$. a is the identity element $\therefore$ each element possesses an inverse which is a member of S. Assuming that $*$ is associative the set S forms a group under $*$.

(iii) The identity elements 0 and a are found in corresponding positions in the tables, similarly each inverse. a corresponds to 0, b to 1, c to 2, d to 3. The groups are isomorphic.

It is perhaps worth noting that these are examples of Abelian groups. When corresponding rows and columns are interchanged in each table, the table is unaltered; the commutative law holds for each operation.

In general, all modulo addition forms a group. Modulo multiplication forms a closed set, the identity element is always 1, the associative and commutative laws hold, but not every element has an inverse so it does not form a group. Note that it is possible for a set to be a group under one operation but not under another.

Key terms

Digits 0, 1, 2...9 the figures used to form our numbers.
Real numbers are all the numbers on the number line. Integers, whole numbers, natural numbers, rational and irrational numbers and prime numbers are all real numbers.
Denary numbers are the ordinary numbers, i.e. base 10 numbers.
Binary numbers are numbers expressed to base 2.
Bicimals are used to express fractions in binary form.
Octal numbers are numbers expressed to base 8.
Modulo or clock arithmetic supposes that only certain digits exist and these cannot be combined. In modulo 4 only $\{0, 1, 2, 3\}$ exist. A complete arithmetic can be built with unique results. E.g. $2+3 = 1$. The clock method is the most convenient way of producing the results.
A **binary operation** is an operation carried out on two numbers. The symbol $*$ is used. It can denote $+, -, \times, \div$, or an unusual operation. E.g. $a*b$ is the result of adding 5 to $3a+b$.
A set S is **closed** under operation $*$ if, for all elements a and b in S, $a*b$ is also in S.
S will have an **identity** element i if for every element $a \in S$ $a*i = i*a = a$.
The operation $*$ is **associative** if, for every $a, b, c \in S$ $(a*b)*c = a*(b*c)$.
The **inverse** of a, written a^{-1}, is such that $a*a^{-1} = a^{-1}*a = i$ (the identity element).
A set forms a **group** under the operation $*$ if it is closed, associative, has an identity element and each element $a \in S$ has an inverse $a^{-1} \in S$.
Isomorphic groups have the same structure. The tables formed under their different operations show the same features.

Chapter 4
Functions (1)

A **relation** is a connection between members of two sets. E.g. in the statement '2 is a prime factor of 8', 'is a prime factor of' is the relation. Figure 8(a) shows the relation for the numbers $\{2, 3, 5\}$ called the **domain**, on a set $\{4, 5, 6, 8\}$ called the **range**. This relation is known as a **mapping**, because each member of the domain maps on to one or more members of the range. Notice that all the arrows lead from the domain to the range.

A relation is a **function** when only one arrow leaves each member of the domain. Figure 8(b) shows the relation 'shoe size' for three men. People normally take only one size of shoe, so only one arrow applies to each name.

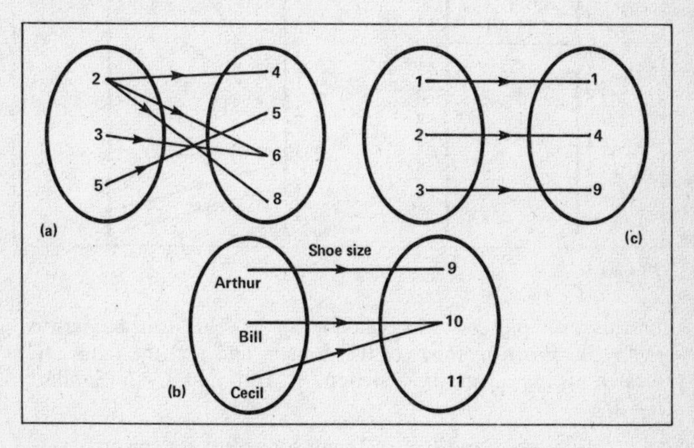

Figure 8

In the case of the relation 'is the square of', shown in figure 8(c), one arrow leaves each member of the domain to go to each member of the range. This function has **one-to-one correspondence**.

Algebraic functions

When we are talking about relations between numbers rather than shoe sizes, it is possible to use symbols. The relation 'is the square of' is written $x \rightarrow x^2$; 'is two more than' is written $x \rightarrow x+2$. Each value of the domain x maps on to one value of the range x^2 or $x+2$, $\therefore$ these relations are functions of x.

The relation 'is the square of' is now written $f: x \to x^2$, where f means function. Similarly $f: x \to x+2$ means that f is the function which maps x on to $x+2$.

$f: x \to 3x$ is the function which maps x on to $3x$. If the domain is $\{-2, -1, 0, 1, 2\}$ the range will be $\{-6, -3, 0, 3, 6\}$. Figure 9(a) shows what is called the **mapping diagram**. Two number lines are used.

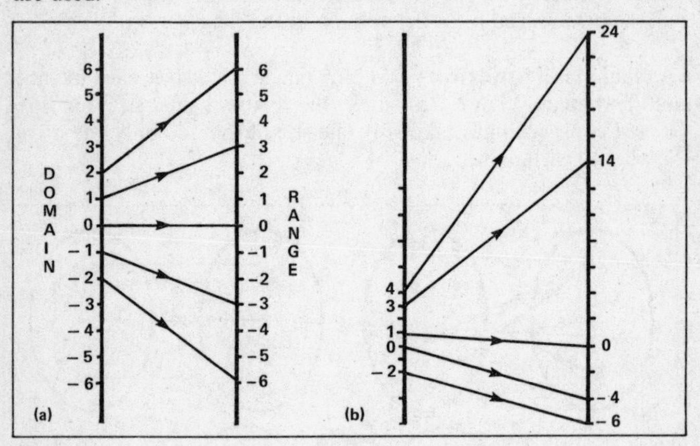

Figure 9

In this example $3x$ is a function of x. This can be written $f(x) = 3x$. x is a member of the domain and $f(x)$ the range. $f(2)$ means 'put $x = 2$ in the function' $\therefore$ $f(2) = 3 \times 2 = 6$. Similarly $f(-2) = -6$.

$f: x \to 3x$ and $f(x) = 3x$ are equivalent statements.

Example For $f: x \to x^2 + 3x - 4$ find the range for the domain $(-2, 0, 1, 3, 4)$. Draw a mapping diagram.

Write $f(x) = x^2 + 3x - 4$, substituting each value in turn:

$f(-2) = 4 - 6 - 4 = -6$ $\qquad$ $f(0) = 0 + 0 - 4 = -4$

$f(1) = 1 + 3 - 4 = 0$ $\qquad$ $f(3) = 9 + 9 - 4 = 14$

$f(4) = 16 + 12 - 4 = 24$

The range is $\{-6, -4, 0, 14, 24\}$. See the mapping diagram in figure 9(b).

Compound functions
$f: x \rightarrow x+2$ and $g: x \rightarrow x^2$ are two functions of x. The **compound function** $f(g(x))$ is abbreviated to fg and says 'carry out g followed by f'. (See page 100 for the equivalent process in geometry.)

If the domain is $\{1, 2, 3\}$ then taking $g(x) = x^2$ first:

$$g(1) = 1 : g(2) = 4 : g(3) = 9$$

We require $f(1) : f(4) : f(9)$ where $f(x) = x+2$

$$f(1) = 1 + 2 = 3 : f(4) = 6 \text{ and } f(9) = 11.$$

$$\therefore \quad fg(1) = 3 : fg(2) = 6 : fg(3) = 11$$

A mapping diagram of the compound function is found in figure 10(a).

In terms of $x : g(x) = x^2 \quad \therefore \quad f(g(x)) = f(x^2)$.

We write x^2 for x in $f(x) = x+2 \quad \therefore \quad f(g(x)) = x^2+2$ (or fg).

Now consider $g(f(x))$ or gf, f followed by g.

$$g(f(x)) = g(x+2) \text{ writing } x + 2 \text{ for } f(x)$$

$$g(x+2) = (x+2)^2 \text{ writing } x + 2 \text{ for } x \quad gf(x) = (x+2)^2$$

We see that $fg \neq gf$. Observe the order right to left carefully. If f, g and h are three functions then fgh means h followed by g followed by f.

Example $f: x \rightarrow x-2$; $g: x \rightarrow 4x$; $h: x \rightarrow x^3$. Evaluate (i) $fg(4)$ (ii) $hfg(-\frac{3}{4})$ (iii) find fgh in terms of x.

Write f, g and h in the form $f(x) = x-2$; $g(x) = 4x$; $h(x) = x^3$.

(i) From right to left: $g(4) = 16, f(16) = 16-2 = 14, \therefore fg(4) = 14$.
(ii) g followed by f followed by h : $g(-\frac{3}{4}) = 4 \times -\frac{3}{4} = -3$.
$f(-3) = -5; h(-5) = (-5)^3 = -125 \quad \therefore \quad hfg(-\frac{3}{4}) = -125$
(iii) $h(x) = x^3$; $g(x^3) = 4x^3$; $f(4x^3) = 4x^3-2$. $\quad \therefore \quad fgh = 4x^3-2$.

Write x^3 for x in g and $4x^3$ for x in f. Verify $hfg(x) = (4x-2)^3$.

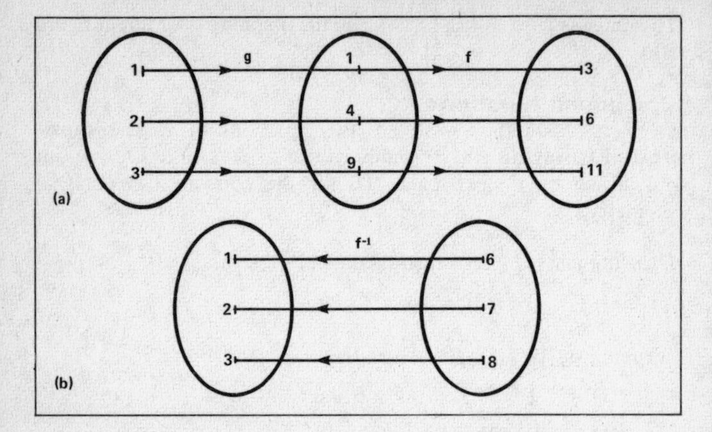

Figure 10

Inverse functions

As already stated, the inverse of addition is subtraction. That of multiplication is division. The inverse of squaring is square rooting. The converse is true in each case.

If $f: x \to x+5$ for the domain $\{1, 2, 3\}$ the range is $\{6, 7, 8\}$. The **inverse of *f***, written $\boldsymbol{f^{-1}}$, is the function which maps the range on to the domain. This is obtained by subtracting 5 from each element. See figure 10(b).

If $f: x \to x+5 \quad f^{-1}: x \to x-5$; if $f: x \to 3x \quad f^{-1}: x \to \dfrac{x}{3}$

If $f: x \to \sqrt{x} \quad f^{-1}: x \to x^2$; if $f: x \to x^2 \quad f^{-1}: x \to +\sqrt{x}$.

In the last of these examples of inverse functions notice the $+\sqrt{x}$. If $x^2 = 9$ then $x = \pm 3$. Two arrows will come from 9 and f^{-1} will not be a function. This is overcome by considering $+\sqrt{x}$ only.

Flow diagrams

Flow diagrams are used to plan the correct sequence of events. They can be used to determine inverse functions of a more complex kind. E.g. for $f: x \to \frac{1}{4}(2x+3)$ a flow diagram can be made of what is done to evaluate f for a given value of x. Work from left to right.

Take x → $\boxed{\times 2}$ → $2x$ → $\boxed{+3}$ → $2x+3$ → $\boxed{\div 4}$ → $\frac{1}{4}(2x+3) = f$

To find f^{-1} take the inverse of every process, from right to left.

$$f^{-1} = \frac{4x-3}{2} \;\leftarrow\; \boxed{\div 2} \;\leftarrow\; 4x-3 \;\leftarrow\; \boxed{-3} \;\leftarrow\; 4x \;\leftarrow\; \boxed{\times 4} \;\leftarrow\; \text{Take } x$$

$\therefore$ if $\quad f : x \to \dfrac{2x+3}{4} \quad$ then $\; f^{-1} : x \to \dfrac{4x-3}{2}$.

Note the following inverses (1) $f : x \to x$ then $f^{-1} : x \to x$;

(2) $f : x \to \dfrac{1}{x}\, f^{-1} : x \to \dfrac{1}{x}$; $\quad$ (3) $ff^{-1}(x) = x$.

To prove (2) make a flow diagram. The inverse of taking the reciprocal is to reciprocate again. f and f^{-1} appear on the same line.

$$\text{Take } x \;\rightarrow\; \boxed{\text{invert}} \;\rightarrow\; \frac{1}{x} = f \qquad\qquad f^{-1} = \frac{1}{x} \;\leftarrow\; \boxed{\text{invert}} \;\leftarrow\; \text{Take } x$$

Verify (1) in a similar way. (3) is verified in part of the next example.

Example $f : x \to \dfrac{2}{x} + 3$. Find (i) $f^{-1}(4)$ (ii) $ff^{-1}(4)$.

Find f^{-1} from a flow diagram:

$$\text{Take } x \;\rightarrow\; \boxed{\text{invert}} \;\rightarrow\; \frac{1}{x} \;\rightarrow\; \boxed{\times 2} \;\rightarrow\; \frac{2}{x} \;\rightarrow\; \boxed{+3} \;\rightarrow\; \frac{2}{x} + 3 = f$$

$$f^{-1} = \frac{2}{x-3} \;\leftarrow\; \boxed{\text{invert}} \;\leftarrow\; \frac{x-3}{2} \;\leftarrow\; \boxed{\div 2} \;\leftarrow\; x-3 \;\leftarrow\; \boxed{-3} \;\leftarrow\; \text{Take } x$$

$\therefore \quad f^{-1}(x) = \dfrac{2}{x-3}$.

(i) $f^{-1}(4) = \dfrac{2}{4-3} = 2$.

(ii) $ff^{-1}(4) = f(2) : f(2) = \dfrac{2}{2} + 3 = 1 + 3 = 4$

$\therefore ff^{-1}(4) = 4$ which verifies the fact that $ff^{-1}(x) = x$.

Example $f: x \to \frac{2}{x}$ and $g: x \to \frac{x}{2}$. Are the following true or false?
(i) $gf = fg$ (ii) $(fg)^{-1} = g^{-1}f^{-1}$.

Find $f^{-1}(x)$ from a flow diagram:

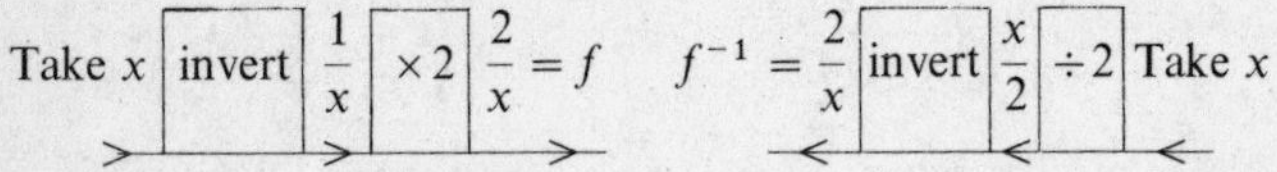

$$\text{Take } x \xrightarrow{} \boxed{\text{invert}} \xrightarrow{} \frac{1}{x} \boxed{\times 2} \xrightarrow{} \frac{2}{x} = f \qquad f^{-1} = \frac{2}{x} \xleftarrow{} \boxed{\text{invert}} \xleftarrow{} \frac{x}{2} \boxed{\div 2} \xleftarrow{} \text{Take } x$$

$\therefore f^{-1}(x) = \frac{2}{x}$ and similarly $g^{-1}(x) = 2x$.

(i) $gf = g(f(x)) = g\left(\frac{2}{x}\right) = \frac{2}{x} \times \frac{1}{2} = \frac{1}{x}$.

$fg = f(g(x)) = f\left(\frac{x}{2}\right) = 2 \div \frac{x}{2} = 2 \times \frac{2}{x} = \frac{4}{x}$ $\therefore$ (i) is false.

(ii) From (i) $fg = \frac{4}{x}$ $(fg)^{-1} = \frac{4}{x}$. The inverse of $\frac{4}{x}$ is also $\frac{4}{x}$.

$g^{-1}f^{-1} = g^{-1}(f^{-1}(x)) = g^{-1}\left(\frac{2}{x}\right) = 2 \times \frac{2}{x} = \frac{4}{x}$.

$\therefore (fg)^{-1} = g^{-1}f^{-1}$ is true.

Rearrangement of formulae

Formulae can be rearranged using flow charts. This is an alternative to that shown on page 26.

To make k the subject of the formula $y = \frac{A(b+k)}{t}$.

The flow diagram evaluates y for values of k, t, b, A. Start with k.

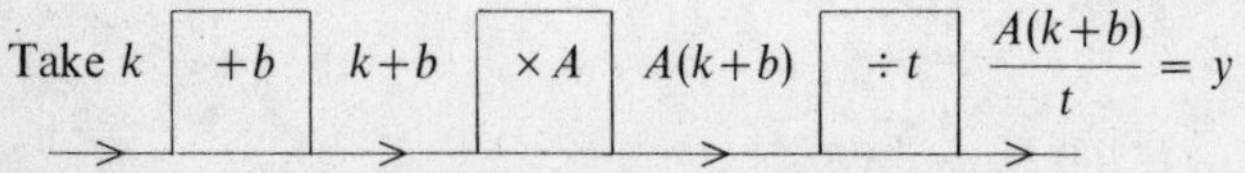

$$\text{Take } k \xrightarrow{} \boxed{+b} \xrightarrow{} k+b \boxed{\times A} \xrightarrow{} A(k+b) \boxed{\div t} \xrightarrow{} \frac{A(k+b)}{t} = y$$

To find k, reverse the order, take inverses, starting with y.

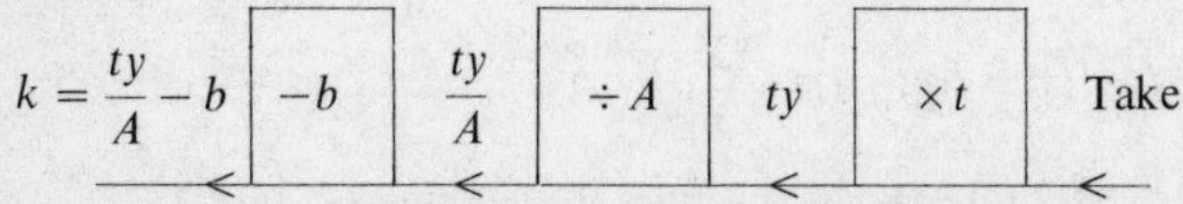

$$k = \frac{ty}{A} - b \xleftarrow{} \boxed{-b} \xleftarrow{} \frac{ty}{A} \boxed{\div A} \xleftarrow{} ty \boxed{\times t} \xleftarrow{} \text{Take } y$$

$$\therefore \quad k = \frac{ty}{A} - b \quad \text{or} \quad \frac{ty - Ab}{A} \quad \text{over one denominator.}$$

Start the first flow chart with the new subject and start the inverse diagram with the old subject.

Example Make l the subject of $T = 2\pi\sqrt{\dfrac{l}{g}}$.

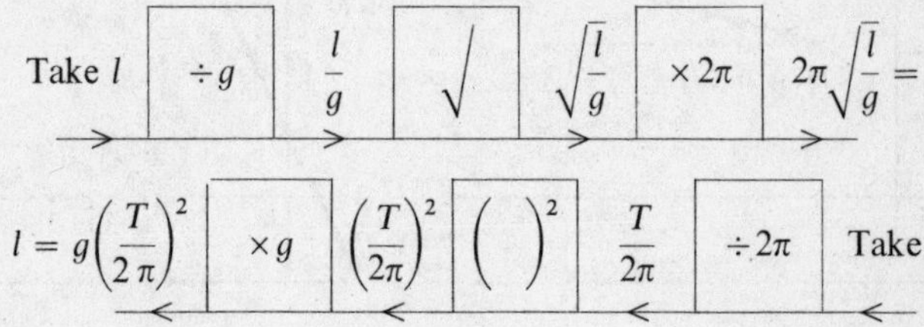

Take l → $\boxed{\div g}$ → $\dfrac{l}{g}$ → $\boxed{\sqrt{}}$ → $\sqrt{\dfrac{l}{g}}$ → $\boxed{\times 2\pi}$ → $2\pi\sqrt{\dfrac{l}{g}} = T$

$l = g\left(\dfrac{T}{2\pi}\right)^2$ ← $\boxed{\times g}$ ← $\left(\dfrac{T}{2\pi}\right)^2$ ← $\boxed{(\ )^2}$ ← $\dfrac{T}{2\pi}$ ← $\boxed{\div 2\pi}$ ← Take T

$$\therefore \quad l = g\left(\frac{T}{2\pi}\right)^2 \quad \text{or} \quad g\frac{T^2}{4\pi^2}$$

Co-ordinates

The co-ordinates of a point in a plane are (x, y), where x is the distance from the vertical y axis and y is the distance from the horizontal x axis. Hence in figure 11(a) the point Q (5, 3) is 5 units from the y axis and 3 units from the x axis.

A point on the x axis will be of the form $(x, 0)$.
A point on the y axis will be of the form $(0, y)$.
The point $(0, 0)$ is called the origin.

To find the distance between the points $P(1, 2)$ and $Q(5, 3)$, complete the right-angled triangle PQR in figure 11(a).
$PR = 5 - 1 = 4$ units, the difference between the x values.
$QR = 3 - 2 = 1$ unit, the difference between the y values.
Using Pythagoras' theorem, $PQ = \sqrt{4^2 + 1^2} = \sqrt{17} = 4.12$ units.

To find the **slope** of the line PQ in the same figure:

$$\text{the slope, or } \textbf{gradient,} \text{ of } PQ = \frac{\text{vertical distance}}{\text{horizontal distance}} = \frac{QR}{PR} = \frac{1}{4}$$

This is **tan *QPR*** in $\triangle PQR$.

$$\text{In general the slope} = \frac{\text{the difference of the } y \text{ values}}{\text{the difference of the } x \text{ values}}$$

If the line slopes 'backwards' the value of the slope is negative.

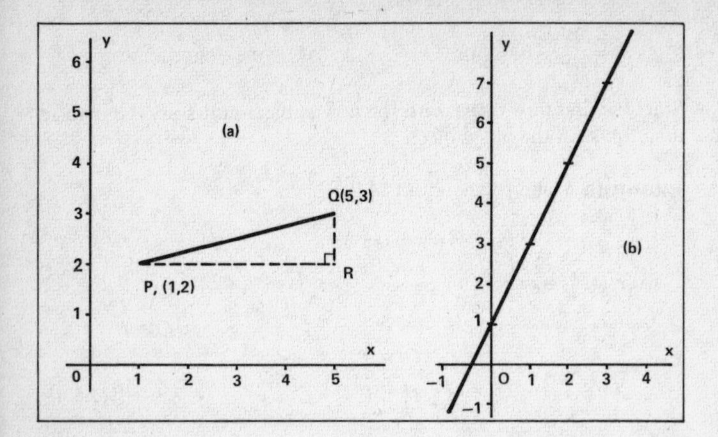

Figure 11

The straight line

If $f: x \to 2x+1$ or $f(x) = 2x+1$ and the domain is $\{0, 1, 2, 3\}$ the range of $f(x)$ is $1, 3, 5, 7$. 0 maps on to 1, 2 maps on to 5, etc. Pair these as co-ordinates $(0, 1)$, $(1, 3)$, $(2, 5)$, $(3, 7)$ and they can be plotted on a graph, taking the domain along the x axis and the range along the y axis. This means that y is a function of x, or $y = f(x)$ and we can write $y = 2x+1$.

The points are plotted in figure 11(b). They form a **straight line**. Any other point on the line satisfies the equation. E.g. the point $(7, 15)$ is on the line because $15 = 2\times 7+1$.

In general the straight-line graph is of the form $y = mx+c$. It is called the **linear equation**.

When $x = 0$, $y = 0+c$, so the line cuts the y axis at $(0, c)$.

When $x = 1$, $y = m+c$, so in figure 12(a) the co-ordinates of B are $(1, m+c)$. Completing the triangle ABC:

the vertical distance $BC = m+c-c = m$
the horizontal distance $AC = 1-0 = 1$

$\therefore$ the slope of the line $= \frac{m}{1}$ or just m.

So the equation $y = mx+c$ represents the straight line with slope m which cuts the y axis at the point $(0, c)$.

The equation $y = 3$ is a horizontal line through the point (0, 3). The equation $x = 2$ is a vertical line through the point (2, 0). These lines intersect at the point (2, 3). See figure 12(b).

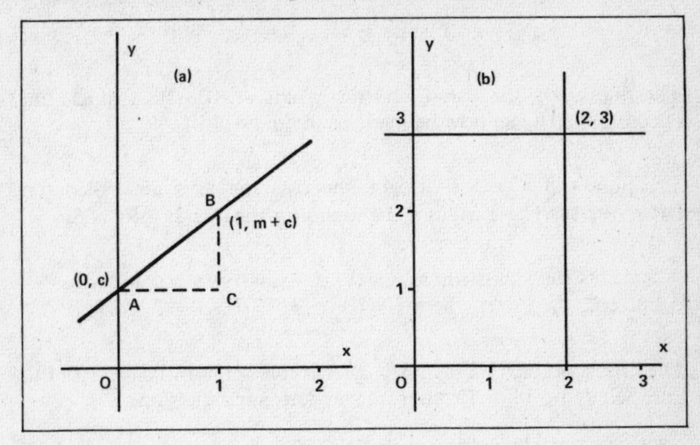

Figure 12

$y = x$ is the line through the origin with gradient 1 $\therefore$ it makes an angle of 45° with both axes. $y = -x$ is the line through the origin, sloping 'backwards' at 45° to both axes. See figure 13(a).

To find the slope and intersection with the y axis for the line $3x - 2y = 4$, rearrange it to the form $y = mx + c$.

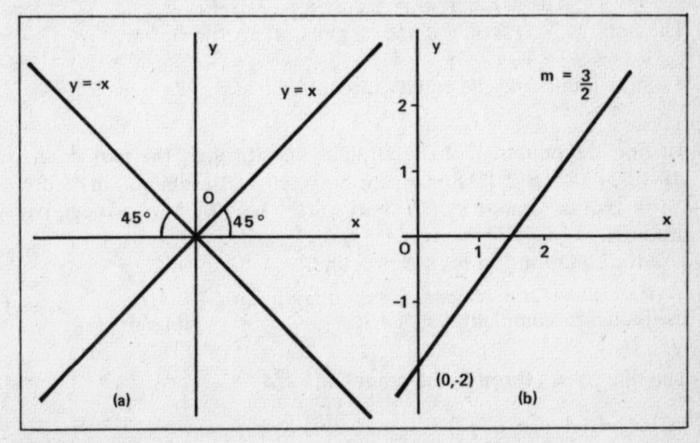

Figure 13

$$3x - 2y = 4 \Leftrightarrow -2y = 4 - 3x \Leftrightarrow y = \frac{4}{-2} - \frac{3x}{-2}$$

$$\Leftrightarrow y = -2 + \frac{3}{2}x \quad \text{or} \quad y = \frac{3}{2}x - 2. \qquad m = \frac{3}{2} \quad \text{and} \quad c = -2.$$

The line has slope $\frac{3}{2}$ and cuts the y axis at $(0, -2)$. A quick but effective sketch can now be made as in figure 13(b).

The points in which a straight line cuts the axes are called the **intercepts**. These will now be found for the line $4x + 3y = 6$.

A line cuts the x axis where $y = 0$. $\quad 4x = 6 \Rightarrow x = 1{\cdot}5$
A line cuts the y axis where $x = 0$. $\quad 3y = 6 \Rightarrow y = 2$

The intercepts are $(1{\cdot}5, 0)$ and $(0, 2)$. A sketch can now be made of the line. See figure 14(a). From it we can determine the slope.

$$\frac{\text{vertical}}{\text{horizontal}} = \frac{2}{1{\cdot}5} = \frac{4}{3} \quad \therefore \quad \text{the slope} = -\frac{4}{3} \text{(– for backwards)}$$

To find the equation of the straight line with gradient -2, which passes through the point $(4, -1)$.

Although we do not know where it cuts the y axis, we use the form $y = mx + c$. The gradient $m = -2 \therefore y = -2x + c$.
The point $(4, -1)$ is on the line $\therefore$ must satisfy the equation.
$\therefore -1 = -2 \times 4 + c \quad$ or $\quad -1 = -8 + c \Rightarrow c = 7$
c is now found and the equation required is $y = -2x + 7$.

To find the equation of the straight line through the two points $A(-3, 5)$ and $B(2, 1)$, first make a sketch of the line as in figure 14(b). Notice that it slopes 'backwards' so will have a negative gradient.
Use the equation $y = mx + c$ as before.

In the figure complete the $\triangle ABC$. $\quad \frac{AC}{BC} = \frac{4}{5} \therefore$ slope $m = -\frac{4}{5}$

The line passes through the point $(2, 1) \therefore 1 = -\frac{4}{5} \times 2 + c$

$\Rightarrow 1 = -\frac{8}{5} + c \Rightarrow c = \frac{8}{5} + 1 \Rightarrow c = \frac{13}{5}$

$\therefore$ the required equation is $y = -\frac{4}{5}x + \frac{13}{5} \quad$ or $\quad 5y + 4x = 13$.

To find the equation of a line parallel to the line $3x-2y+4=0$ which passes through the point $(4,-2)$.

Parallel lines have the same gradient $\therefore$ the x and y terms of the required line can be written as those of the given line.
$\therefore$ the new line will be $3x-2y+k=0$ where k is a constant which can be found if one point on the line is known.

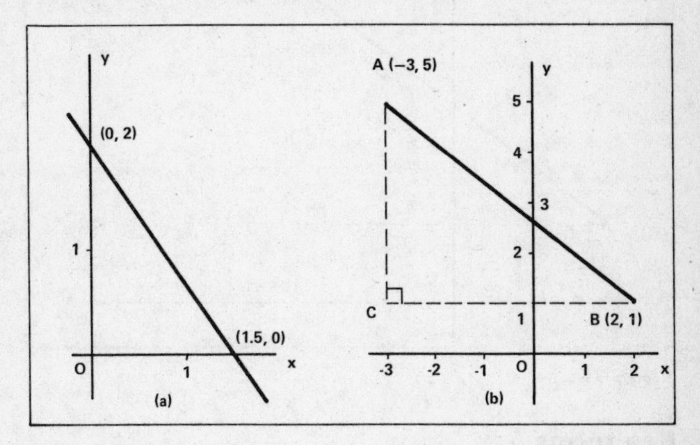

Figure 14

It passes through $(4,-2) \therefore 12-2(-2)+k=0 \Rightarrow 16+k=0 \Rightarrow$ $k=-16$ and the line is $3x-2y-16=0$.

Example A and B are the points $(-2,1)$ and $(2,4)$ respectively. Find (i) the gradient of AB; (ii) the length of AB; (iii) the co-ordinates of the mid-point of AB; (iv) the equation of the line through $(4,5)$ which is parallel to AB; (v) the equation of line OB.

(i) Make a sketch as in figure 15. Complete the $\triangle ABC$. The gradient of $AB = \dfrac{BC}{AC} = \dfrac{3}{4}$ (positive)

(ii) Using Pythagoras' theorem, $AB = \sqrt{3^2+4^2} = \sqrt{25} = 5$ (a 3:4:5 triangle).

(iii) To find the point half-way between A and B find the average of the co-ordinates. $\left(\dfrac{-2+2}{2}, \dfrac{1+4}{2}\right) = (0, 2\frac{1}{2})$

(iv) Let the line be $y = mx + c$. From (i) $m = \frac{3}{4}$ $\therefore$ $y = \frac{3}{4}x + c$.
The required equation is $y = \frac{3}{4}x + 2$ or $4y = 3x + 8$.

(v) OB passes through $(0, 0)$ and $(2, 4)$ $\therefore$ $c = 0$ and $m = \frac{4}{2} = 2$.
The required equation is $y = 2x$.

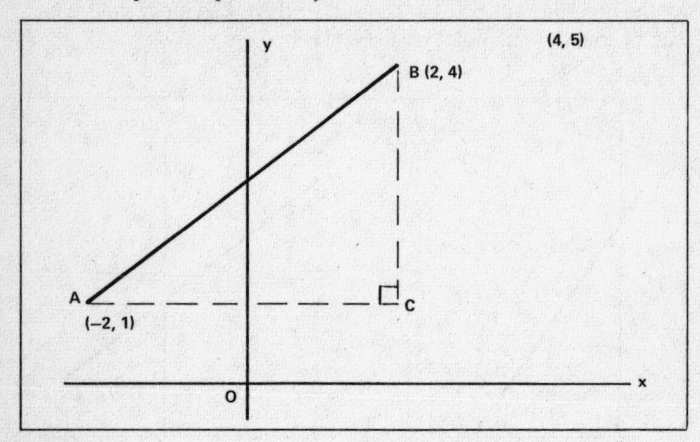

Figure 15

Key terms

A **relation** is the connection between two sets A and B.
A **mapping** is the linking of elements of A with their images in B.
If the elements of A are mapped on to those of B, A is called the **domain** and B the **range**.
A relation is a **function** if each element of the domain has only one image in the range, even if some of the elements of the range are not used.
A function has a **one-to-one correspondence** if every point of A is mapped on to a unique element in B and no two points of A are mapped on to the same element of B. See figure 8, page 41.
A **compound** function $fg(x)$ means g followed by f.
The **inverse** function of f, f^{-1}, maps the range on to the domain.
The **gradient** of a line $= \dfrac{\text{vertical}}{\text{horizontal}} =$ the tangent of the angle.
The **linear** equation is the equation of a straight line. We consider the form $y = mx + c$.
The **intercepts** of a straight line are the points at which it cuts the two axes.

Chapter 5
Functions (2)

If $f(x) = x^2$, for the domain $\{-3, -2, -1, 0, 1, 2, 3\}$ the range of $f(x)$ is $\{9, 4, 1, 0, 1, 4, 9\}$. In figure 16(a), $f(x)$ or y is plotted against x. y is positive for all values of x. Figure 16(b) shows the graph of $y = x^2 - 2$; each value of y is reduced by 2. Figure 16(c) shows the graph of $y = -x^2$ the reflection in the x axis of $y = x^2$. Figure 16(d) shows $y = 2 - x^2$ the reflection of $y = x^2 - 2$ in the x axis. This curve is called a **parabola**. It has an **axis of symmetry** which is the y axis in these examples.

The graph of the quadratic function $f(x) = ax^2 + bx + c$ is a parabola.

To sketch the curve $y = x^2 - 3x + 2$, find its intersections with the axes.

On the y axis $x = 0 \therefore y = 2$. It cuts the y axis at (0, 2).

On the x axis $y = 0 \therefore 0 = x^2 - 3x + 2$, a quadratic equation. Solving

$\Leftrightarrow (x-1)(x-2) = 0 \Leftrightarrow x - 1 = 0$ or $x - 2 = 0$

$\Leftrightarrow x = 1$ or $x = 2$.

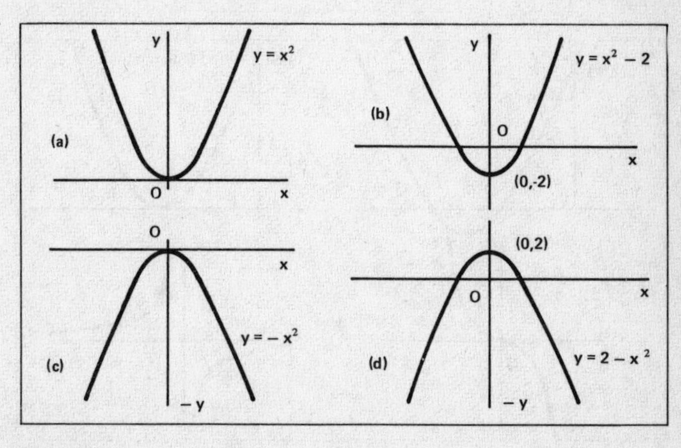

Figure 16

It cuts the x axis at $(1,0)$ and $(2,0)$. The curve can now be sketched as in figure 17(a), joining the three points in a parabola.

The curve $y = x^2 - 6x + 9$ cuts the y axis at $(0,9)$, putting $x = 0$. When $y = 0$

$$0 = x^2 - 6x + 9 \Leftrightarrow (x-3)(x-3) = 0 \Leftrightarrow x = 3 \text{ only.}$$

The curve meets the x axis in one point. The axis is a tangent to the curve. See figure 17(b).

Figure 17(c) shows the graph of $y = x^3$; y is negative when x is and y increases as the cube of x numerically.

Figure 17(d) shows the graph of $y = \dfrac{1}{x}$, the hyperbola. For the domain $\{-3, -2, -1, 0, 1, 2, 3\}$ the range is $\{-\frac{1}{3}, -\frac{1}{2}, -1, \infty, 1, \frac{1}{2}, \frac{1}{3}\}$. y is the reciprocal of x so decreases numerically as x increases. When $x = 0$, $y = \frac{1}{0}$ which is ∞. The graph nears the y axis as x tends to 0.

Variation

In the table it can be seen that the value of y is 3 times the

x	1	2	3	4	5
y	3	6	9	12	15

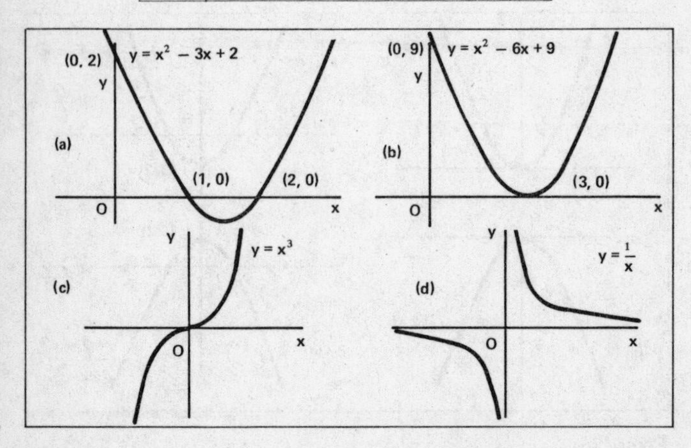

Figure 17

value of x. The value of y depends directly on the value of x. It is said that 'y **varies directly** as x' or just 'y varies as x'. It is written $y \propto x$ or $\mathbf{y = kx}$, where k is a constant.

$y = kx \Leftrightarrow \frac{y}{x} = k$ i.e. the ratio $y : x$ is constant, so it can also be said that 'y is **directly proportional** to x'. The graph of 'y varies as x' is a straight line through the origin with gradient k.

Example If y varies as x and $x = 4$ when $y = 20$, find the value of y when $x = 56$.

'y varies as x' means that $y = kx$. $x = 4$ when $y = 20$ $\therefore$ $20 = 4k$ $\Rightarrow k = 5$. The relationship between y and x is $y = 5x$. When $x = 56$ $y = 5 \times 56$ $\therefore$ $y = 280$.

The circumference of a circle varies as the radius, because $C = 2\pi r$ i.e. $C = kr$ where $k = 2\pi$.

In the table each value of y is twice the square of x i.e. $y = 2x^2$.

x	0	1	2	3	4	5
y	0	2	8	18	32	50

In this case 'y **varies** as the **square** of x'. This is written $y \propto x^2$ or $y = kx^2$. It is also said that 'y is directly proportional to the square of x'. The graph of $y = kx^2$ is a parabola through the origin, each value of k giving a different parabola. Given one particular point on the curve, k can be found and the required curve selected.

Example If y varies as the square of x and $x = 4$ when $y = 48$, find y when $x = 7$.

We know that $y = kx^2$ and that the point $(4, 48)$ lies on the curve. $\therefore$ $48 = k \times 4^2 \Leftrightarrow 16k = 48 \Leftrightarrow k = 3$. The required curve is $y = 3x^2$. When $x = 7$, $y = 3 \times 49 \Leftrightarrow y = 147$.

The area of a circle varies as the square of the radius because $A = \pi r^2$, i.e. $A = kr^2$ where $k = \pi$. Similarly, the surface area of a sphere varies as the square of the radius, because $A = 4\pi r^2$.

If y **varies** as the **cube** of x use the relationship $y = kx^3$.

In this table the value of y is obtained by taking twice the inverse

x	1	2	3	4	5	6
y	2	1	$\frac{2}{3}$	$\frac{1}{2}$	$\frac{2}{5}$	$\frac{1}{3}$

of x. 'y **varies inversely** as x' or 'y is **inversely proportional** to x'. It is written $y \propto \frac{1}{x}$ or $y = \frac{k}{x}$. The graph of the function is a hyperbola as in figure 17(d). Different values of k give different curves. Given a point on the curve k can be determined.

Example If y varies inversely as x and $y = 8$ when $x = 30$, find the value of x when $y = 60$.

We know that $y = \frac{k}{x}$ and $x = 30$ when $y = 8$.

$\therefore \quad 8 = \frac{k}{30} \Leftrightarrow k = 240.$

$\therefore$ the required relationship is $\quad y = \frac{240}{x}$.

When $y = 60$, $\quad 60 = \frac{240}{x} \Rightarrow x = \frac{240}{60} \Rightarrow x = 4.$

Boyle's law states that the pressure p of a given mass of gas at constant temperature varies inversely as the volume of the gas, $\therefore p = \frac{k}{v}$. The area of a rectangle $A = lb \Leftrightarrow b = \frac{A}{l}$. The breadth of the rectangle varies inversely as the length, if the area is constant.

If y varies **inversely** as the **square** of x then we write $y = \frac{k}{x^2}$.

The formula for the volume of a cylinder is $V = \pi r^2 h$. The volume depends on the radius squared and the height at the same time. It is said that 'V **varies** as the square of r and as h **jointly**'.

Example y varies as x and the inverse of z jointly. Complete the given table

x	3	9	12
y	5		40
z	4	12	

The relationship is $y = \frac{kx}{z}$. Use $x = 3$, $y = 5$, $z = 4$ to find k.

$$5 = k \times \tfrac{3}{4} \Leftrightarrow 3k = 20 \Leftrightarrow k = \frac{20}{3} \quad \therefore \quad y = \frac{20x}{3z}$$

Complete the table using the other values given:

first $x = 12$, $y = 40$. $40 = \frac{20 \times 12}{3z} \Leftrightarrow 3z = \frac{240}{40} \Leftrightarrow z = 2$

now $x = 9$, $z = 12$. $y = \frac{20 \times 9}{3 \times 12} \Leftrightarrow y = 5$

and the table is complete.

The straight-line law

If y varies as $+\sqrt{x}$ then $y = k\sqrt{x}$. If $k = 2$ then $y = 2\sqrt{x}$. The table shows some values of y for given values of x.

x	0	1	2	3	4	5	6
y	0	2	2·828	3·464	4	4·472	4·90
$\sqrt{x}$	0	1	1·414	1·732	2	2·236	2·449

To plot the graph this time take the horizontal axis as $\sqrt{x}$, keeping y as the vertical axis. This is called **plotting** y **against** $\sqrt{x}$. The result appears in figure 18(a). It is a straight line with gradient 2 (the value of k). This will occur with all the variation relationships. Verify that if $y = 3x^2$ and y is plotted against x^2 the graph will be a **straight line** of gradient 3. By taking the horizontal axis as x^2 instead of x, we have let $x^2 = X$. The graph plotted is $y = 3X$ which is linear.

Rates of change

Figure 18(b) shows the distance (in km) travelled by a car, against time (in hours). Between 0900 hrs and 1000 hrs the car travels 40 km, i.e. its speed is 40 km/hr. The line OA gives this part of the journey.

$$\text{Its slope} = \frac{\text{vertical}}{\text{horizontal}} = \frac{\text{distance}}{\text{time}} = \text{speed} = 40\,\text{km/hr.}$$

The **gradient** denotes the **speed** or **rate of change** of distance.

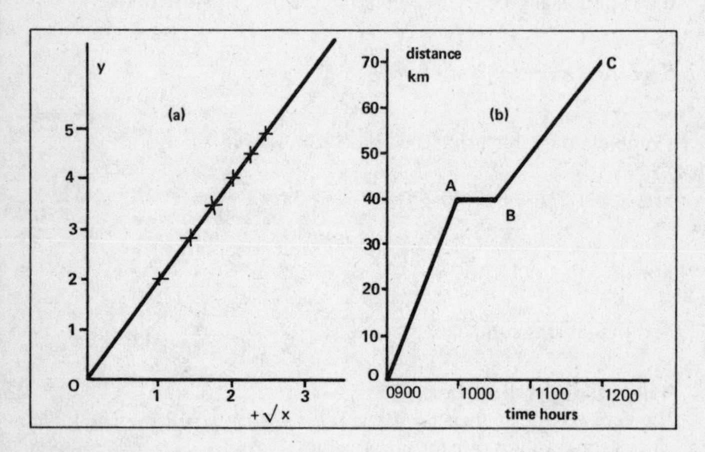

Figure 18

The line AB shows the car standing still for $\frac{1}{2}$ hr. There is no addition to the distance. A zero gradient gives a zero speed. Between 1030 and 1200 hrs the car travels 30 km $\therefore$ its speed $= 30 \div 1\frac{1}{2} =$ 20 km/hr.

$$\text{The average speed for the whole journey} = \frac{\text{total distance}}{\text{total time}} = \frac{70}{3}$$

$$= 23{\cdot}3\ \text{km/hr.}$$

The graph of $y = x^2 + 1$ for $x \geqslant 0$ is drawn in figure 19(a). The rate of change varies at each point on a curve. To measure it we find the slope of the tangent at the point required, either by plotting the graph accurately as shown on page 61, or by calculus (see page 176).

On a **velocity/time** graph the **gradient** $= \dfrac{\textbf{velocity}}{\textbf{time}}$

$= \textbf{acceleration}$

E.g. In figure 19(b) the line OA shows an object starting at rest and reaching a speed of 9 m/s two seconds later. Its acceleration $= 9 \div 2 = 4{\cdot}5$ m per second per second (written m/s^2). The horizontal line AB shows that the speed remains constant at 9 m/s. A zero gradient gives a zero acceleration. BC with its negative slope shows the object coming to rest in 4 seconds, with an acceleration of $-9 \div 4 = -2{\cdot}25$ m/s^2, i.e. deceleration. The slope of the curve $y = f(x)$ measures the rate of change of y.

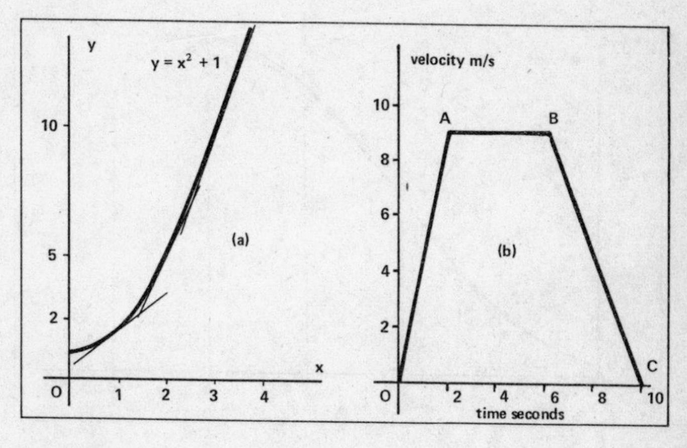

Figure 19

The area under the velocity/time graph = the distance travelled.

In trapezium $OABC$, $AB = 4$ seconds, $OC = 10$ seconds. Its height = 9 m/s. The area $= \frac{1}{2}(a+b)h = \frac{1}{2}(4+10) \times 9 = 63$.

We have multiplied speed (vertical) by time (horizontal), to find the distance. $\therefore$ the area 63 units^2 = the distance travelled 63 m.

Figure 20 shows the velocity/time graph $v = 1 + 6t^2 - t^3$ m/s. When $t = 0$ $v = 1$. The initial velocity = 1 m/s. Between $t = 0$ and $t = 2$

the gradient increases, i.e. the acceleration increases, to its maximum at $t = 2$. Thereafter the acceleration decreases until $t = 4$ when it is zero. At this point the velocity is maximum 33 m/s. After $t = 4$ the negative slope denotes deceleration.

To estimate the maximum acceleration, draw the tangent at $t = 2$. On the tangent the velocity increases 12 units as t goes from 1 to 2 $\therefore$ the slope $= \frac{12}{1}$. The maximum acceleration $=$ 12m/second^2.

The trapezium rule

To find an approximate area under the curve in figure 20, divide it up into **trapeziums**, taking verticals through $t = 0, 1 \ldots 5$, and assuming the sections of the curve cut off to be straight lines. Substituting $t = 0, 1 \ldots 5$ to find the vertical height v in each case:

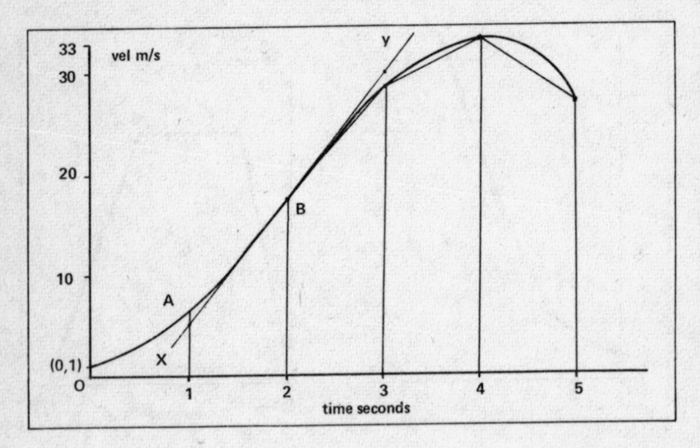

Figure 20

When $t = 0$ $v = 1$; $t = 1$ $v = 1+6-1 = 6$; $t = 2$ $v = 1+24-8 = 17$; $t = 3$ $v = 28$; $t = 4$ $v = 33$ and $t = 5$ $v = 26$.
The approximate area under the curve between $t = 0$ and $t = 5$ is:

$$\frac{1+6}{2} \times 1 + \frac{6+17}{2} \times 1 + \frac{17+28}{2} \times 1 + \frac{28+33}{2} \times 1 + \frac{33+26}{2} \times 1$$

$$= 3.5 + 11.5 + 22.5 + 30.5 + 29.5 = 97.5$$

$\therefore$ using trapeziums the distance travelled is 97.5 m approximately.

If the horizontal interval is h and the heights $y_0, y_1, y_2 \cdots y_n$

Area under curve $= \frac{1}{2}h(y_0 + y_n + 2(y_1 + y_2 + y_3 \ldots y_{n-1}))$.

The smaller the value of h the more accurate is the area.

Plotting graphs accurately

By plotting graphs accurately we can solve equations, measure rates of change and study the nature of a function. The more accurate the graph the more precise the answers. Make a clear table of values. If a straight line is to be plotted, three points are sufficient. Note the scale, which is usually given in examination questions. This facilitates marking and has been chosen to produce the best results. The following example involves most of the techniques needed.

Example Plot the graph of the function $y = \frac{1}{2}(6 + 7x - x^3)$ for $x = -2{\cdot}5, -2, -1{\cdot}5, -1, 0, 1, 1{\cdot}5, 2, 2{\cdot}5, 3$ (the domain). Use the graph to find (i) the maximum and minimum values of y; (ii) the roots of the equation $6 + 7x - x^3 = 0$; (iii) the roots of $6 + 7x - x^3 = 2$; (iv) the range of values of x for which $6 + 7x - x^3 \leqslant 0$; (v) the rate of change of y with respect to x at $x = 1$; (vi) by drawing a straight line solve $6 + 7x - x^3 = x + 4$.

x	$-2{\cdot}5$	-2	$-1{\cdot}5$	-1	0	1	$1{\cdot}5$	2	$2{\cdot}5$	3
6	6	6	6	6	6	6	6	6	6	6
$+7x$	$-17\frac{1}{2}$	-14	$-10\frac{1}{2}$	-7	0	7	$10\frac{1}{2}$	14	$17\frac{1}{2}$	21
$-x^3$	$15\frac{5}{8}$	8	$3\frac{3}{8}$	1	0	-1	$-3\frac{3}{8}$	-8	$-15\frac{5}{8}$	-27
Add	$4\frac{1}{8}$	0	$-1\frac{1}{8}$	0	6	12	$13\frac{1}{8}$	12	$7\frac{7}{8}$	0
$\div 2$	$2{\cdot}06$	0	$-0{\cdot}56$	0	3	6	$6{\cdot}56$	6	$3{\cdot}94$	0

Draw up the table of values. Note how it is planned. Add the numbers in each column and then divide them all by 2 (excluding x itself). The results are given in decimal form to facilitate the plotting on metric graph paper. Now look at figure 21.

Plan the axes, so that all points can be plotted. The x axis is placed to allow 6·56 units above and 0·56 units below. The y axis is placed to allow for the range $-2{\cdot}5$ to $+3$ along the x axis.

Mark the points and join them with a smooth continuous curve. Label the axes and the curve. Answer the questions listed above.

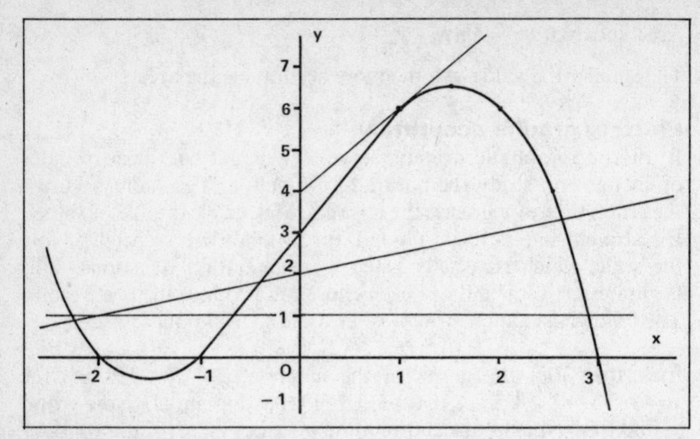

Figure 21

(i) The **maximum** value is 6·56, the **minimum** −0·56. They occur at the turning points approximately (1·53, 6·56) (−1·53, −0·56) respectively.

(ii) $6+7x-x^3=0$ when $y=0$ i.e. where the curve cuts the x axis. These points have exact co-ordinates in this case. $x=3,-2,-1$.

(iii) $6+7x-x^3=2$. The left-hand side is twice the function drawn. Dividing through by 2 gives $\frac{1}{2}(6+7x-x^3)=1$; we require the values of x for which $y=1$. Draw the line $y=1$ to find that it cuts the curve at approximately $x=2{\cdot}9, -0{\cdot}6$, and $-2{\cdot}3$.

(iv) For $6+7x-x^3 \leqslant 0$ we require $y \leqslant 0$, i.e. that part of the curve on or below the x axis. This is when $x \geqslant 3$ and $-1 \geqslant x \geqslant -2$.

(v) Draw the tangent at $x=1$ as accurately as possible. On this tangent, between $x=0$ and $x=1\frac{1}{2}$, y increases from 4 to 7.

$$\therefore \text{ the slope } = \frac{7-4}{1\frac{1}{2}-0} = \frac{3}{1\frac{1}{2}} = 2.$$

The rate of change of y at $x=1$ is 2.

(vi) Arrange the equation so that the left-hand side is the function drawn. $\frac{1}{2}(6+7x-x^3)=\frac{1}{2}(x+4)$ dividing through by 2.

The line to be drawn is $y=\frac{1}{2}(x+4)$. Make a table of values:

x	−2	0	2
$x+4$	2	4	6
÷2	1	2	3

The line cuts the curve at $x = 2{\cdot}6, -0{\cdot}34, -2{\cdot}24$ approximately. These are the required roots of the equation.

Key terms

Parabola is the name given to a graph of the quadratic function ax^2+bx+c.
Variation is the study of the behaviour of one variable as another varies, e.g. directly, inversely, as the square . . .
The straight-line law If the function $y = ax^n$ is plotted with y against x^n the graph is a straight line with gradient a.
The rate of change of y with respect to x is measured by the gradient of the curve at the point (x, y).
Speed is the rate of change of distance with respect to time. In this context it is often called **velocity**. We are not involved with vectors here so there is no need to distinguish between a scalar and a vector. The units of speed are usually metres per second or kilometres per hour abbreviated to m/s or km/h.
Acceleration is the rate of change of velocity with respect to time. The units of measurement are usually metres per second per second or kilometres per hour per hour abbreviated to m/s^2 or km/h^2.
The trapezium rule enables us to find the approximate area under a curve, by dividing the area into a series of convenient trapeziums.

Chapter 6
Inequalities

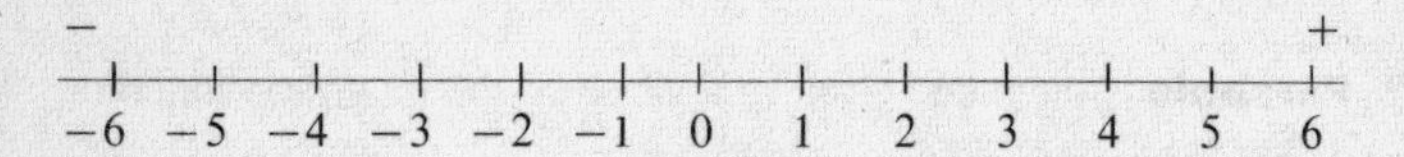

$x > 2$ means that x is any real number on the number line to the right of $x = 2$. $x \leqslant 1$ means that x is any real number **on** or to the left of $x = -1$. $2 \geqslant x > -1$ represents the set of real numbers between $x = 2$ and $x = -1$, **including** $x = 2$.

Inequations

$9 > 3$ is a true statement. It remains true if the same number is added or subtracted from both sides. E.g. Adding 5 gives $14 > 8$. It remains true if multiplied or divided by the same positive number, e.g. multiplying by $+5$ gives $45 > 15$. But when multiplied by -5 we obtain $-45 > -15$, which is false. However it is made true by reversing the arrow. $-45 < -15$.

We can add, subtract, multiply or divide both sides of an inequation by the same number, reversing the arrow when multiplying or dividing by a negative number. This means that inequations can be treated in much the same way as equations.

To solve $4x+9 > 25 \Leftrightarrow 4x > 25-9 \Leftrightarrow 4x > 16$.
$\Leftrightarrow x > 4$ called the **solution set**. We subtract 9 from both sides, then divide through by 4.

To solve $-4x > 18 \Leftrightarrow x < \dfrac{18}{-4} \Leftrightarrow x < -4{\cdot}5$

Divide through by -4, and **reverse** the arrow.

To solve $3(x+4)-2(x+3) > 6x-2$

$\Leftrightarrow 3x + 12 - 2x - 6 > 6x - 2$ removing the brackets.

$\Leftrightarrow 5x+6 > 6x-2 \Leftrightarrow 5x-6x > -2-6 \Leftrightarrow -x > -8 \Leftrightarrow x < 8$.
Rearranging, we find $-x$ on the left. Change the sign and arrow.

To solve $x^2-5x+4 < 0$ make a sketch of the curve $y = x^2-5x+4$ using the method shown on page 53. It cuts the y axis at $(0, 4)$ and when $y = 0$ we have

$x^2 - 5x + 4 = 0 \Leftrightarrow (x-4)(x-1) = 0 \Leftrightarrow x = 4$ or 1.

It cuts the x axis at (1,0) and (4,0), the graph of this quadratic being drawn in figure 22(a). The function is less than 0 below the x axis, between the x values 1 and 4, $\therefore$ the solution is $4 > x > 1$.

To solve $x^2 > 4$. Write it as $x^2 - 4 > 0$ and proceed as before, sketching the graph of $y = x^2 - 4$. It cuts the y axis at $(0, -4)$. When $y = 0$, $x^2 - 4 = 0 \Leftrightarrow (x+2)(x-2) = 0 \Leftrightarrow x = \pm 2$.

The curve cuts the x axis at $(-2, 0)$ and $(2, 0)$. The sketch can now be drawn as in figure 22(b).

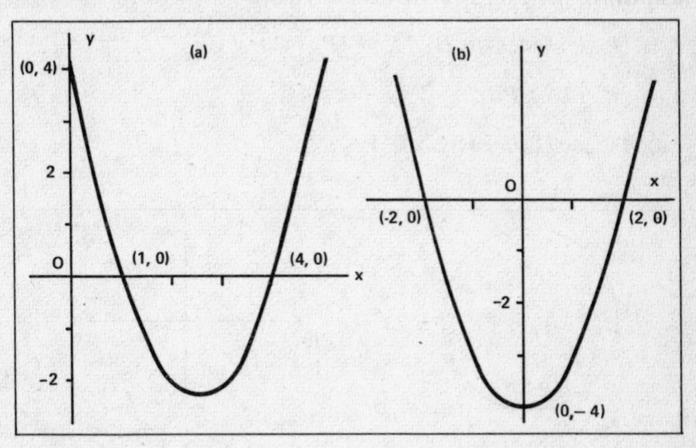

Figure 22

$y > 0$ for the part of the curve above the x axis. This is to the right of $x = 2$ i.e. $x > 2$ and also to the left of $x = -2$ i.e. $x < -2$.

Note that by just taking the square root of both sides the full result would not have been obtained.

Locus

The **locus** of a point is the path or area it traces when forced to move under given conditions. E.g. The locus of a point P which moves so that its distance from a fixed point A is constant r, is a circle of radius r. If the point moves so that the distance from A is less than r, i.e. $AP < r$ then the locus of P is the region inside the circle. See figure 23(a).

Figure 23(b) shows the locus of a point which moves so that it is the same distance from the points A and B. It is the perpendicular bisector of AB, the line XY. The shaded area in figure 23(c) shows the locus of P if P moves so that $AP < PB$. The broken line denotes that P cannot lie on XY.

In figure 23(d) the circles have radii 5 and 3 cm. If a point P is constrained to move in the shaded area, or on it, then the distance AP is less than the distance PB and the distance OP is between 3 and 5 cm and can equal both, $\therefore$ $AP < PB$ and $5 \geqslant OP \geqslant 3$.

Example In figure 23(e) $ABCD$ is a square.

$\mathscr{E} = \{P : P \text{ inside } ABCD\}$ $\quad X = \{P : PA < PC\}$

$Y = \{P : PD < PB\}$

Find the area defined by $X \cap Y$.

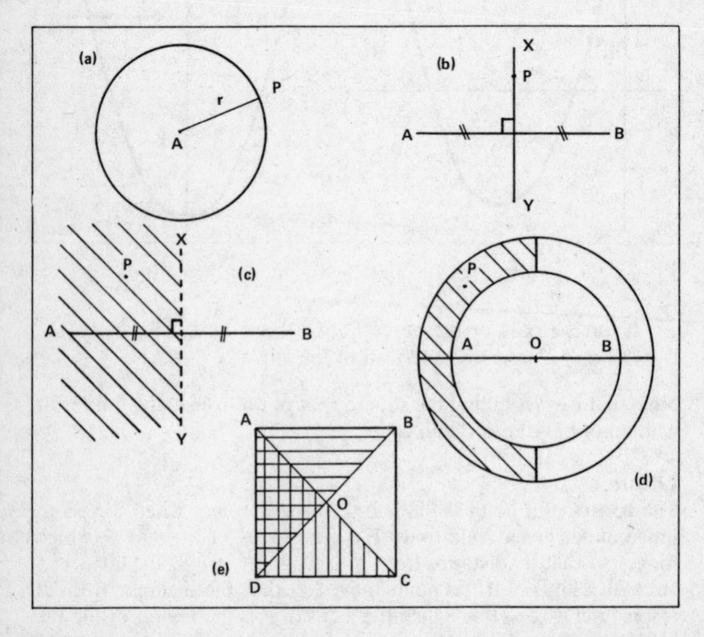

Figure 23

The set X represents points P which are nearer A than C, the points inside the triangle DAB. Shade the inside of this triangle horizontally. Y is the set of points P which are nearer D than B, inside triangle ADC. Shade this triangle vertically. The area defined by $X \cap Y$ is the area shaded both ways. Triangle AOD is the required area or locus.

The co-ordinate plane

This can also be divided into regions defined by inequalities.

In figure 24(a) the shaded area is on or above the x axis, $\therefore\ y \geqslant 0$.
It is to the right of and on the y axis $\therefore\ x \geqslant 0$.
It is on or below the line $y = 2\ \therefore\ y \leqslant 2$.
It is to the left of $x = 3$, but not on the line $\therefore\ x < 3$.
Remember the convention of the broken line.
The region is defined completely by $2 \geqslant y \geqslant 0$ and $3 > x \geqslant 0$.
This is also written $P = \{(x, y): 3 > x \geqslant 0, 2 \geqslant y \geqslant 0\}$.

In figure 24(b) the shaded area is defined by $y \geqslant -x$ and $y \leqslant +x$. Any point in this area lies above the line $y = -x$, or on it. It also

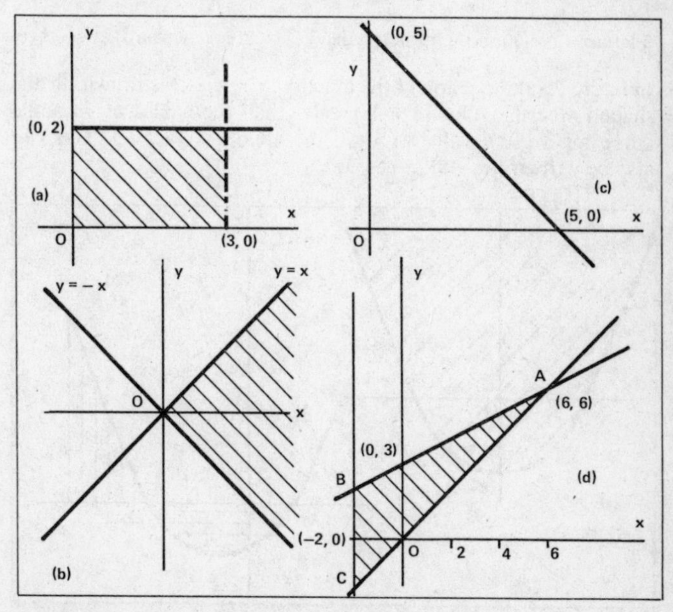

Figure 24

lies on or below the line $y = +x$.

In figure 24(c) the line $x + y = 5$ is drawn.
When $x = 0$ $y = 5$ and when $y = 0$ $x = 5$. The intercepts are $(0, 5)$ and $(5, 0)$.
For any point on the line $x + y = 5$ or $y = 5 - x$.
For a point P above the line $x + y > 5$ or $y > 5 - x$.
For a point Q below the line $x + y < 5$ or $y < 5 - x$.
Again the last statement can be written $Q = \{(x, y): x + y < 5\}$.

To define the area enclosed in figure 24(d), we must find the equations of the lines.
AC passes through the origin and $(6, 6)$ $\therefore$ is the line $y = x$. The shaded area lies on or above this $\therefore$ $y \geqslant x$.

Let AB be of the form $y = mx + c$. It crosses the y axis at $(0, 3)$ $\therefore$ $c = 3$ and $y = mx + 3$. The line also passes through $(6, 6)$ $\therefore$ $6 = 6m + 3 \Leftrightarrow 6m = 3$ $\therefore$ $m = \frac{1}{2}$ and the line is $y = \frac{1}{2}x + 3$.
The shaded area is on or below this line, $\therefore$ $y \leqslant \frac{1}{2}x + 3$ or $2y - x \leqslant 6$
BC is the line $x = -2$, the shaded area is on and to the right of BC $\therefore$ $x \geqslant -2$.
The area is defined completely by $x \geqslant -2$; $y \geqslant x$; and $2y - x \leqslant 6$.

In figure 25(a) the graph of the function $y = x^2 - 3$ is drawn. In the shaded area the value of y is greater than the values of y on the curve for a given value of x, $\therefore$ the area is $y \geqslant x^2 - 3$. This can also be written $\{(x, y): y \geqslant x^2 - 3\}$.

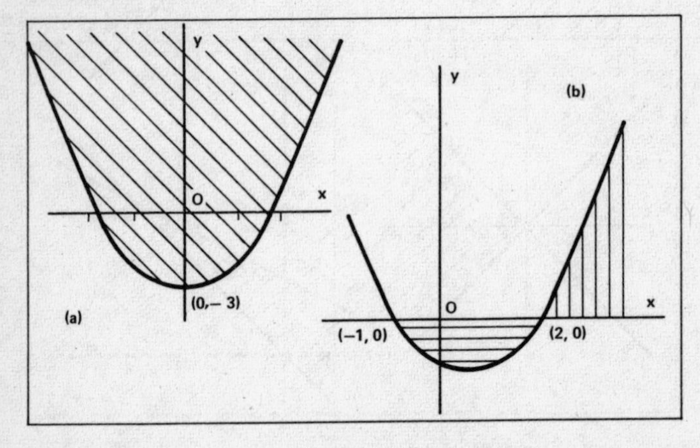

Figure 25

Figure 25(b) shows the graph of the function $y = x^2 - x - 2$. The area shaded horizontally lies below the x axis and above the curve, $y \leqslant 0$ and $y \geqslant x^2 - x - 2$. The vertical shading represents $y \geqslant 0$ and $y \leqslant x^2 - x - 2$.

Example $\mathscr{E} = \{(x, y): x \text{ and } y \text{ real}\}$ on the x, y plane. $A = \{(x, y): x \geqslant 3\}$ $B = \{(x, y): y \leqslant -2\}$ $C = \{(x, y): y \leqslant x\}$. On separate diagrams shade in the areas (i) $A \cap B'$; (ii) C; (ii) $(A \cap B') \cap C$; (iv) in figure 26(d) describe the shaded region in terms of A, B, C.

(i) A is the area on or to the right of $x = 3$, shaded horizontally. B' is the area not less or equal to $y = -2$, i.e. the area $y > -2$. Shade this vertically. (Notice the broken line.) $A \cap B'$ is the area shaded both ways. See figure 26(a).

(ii) In figure 26(b) C is the area on or below the line $y = x$.

(iii) In figure 26(c), shade the area obtained in (i) vertically, shade (ii) horizontally. $(A \cap B') \cap C$ is the area shaded both ways.

(iv) The area contains the points not in A and not in B and in C. The required area is $A' \cap B' \cap C$.

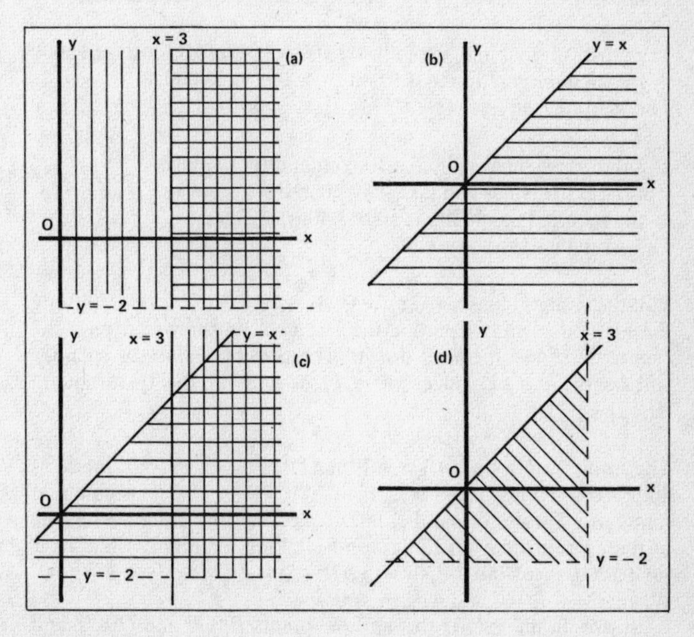

Figure 26

Linear programming
The following example shows how problems involving inequalities can be solved. Only two unknowns are possible using the x, y plane.

Example A factory makes two kinds of machine, A and B. A costs £3 000 to make, B costs £4 000. A requires 100 man days of labour to make and B 400 days. In one specific week the factory has up to 2 000 man days and up to £36 000 in cash available for the manufacture of the machines. The factory must make at least two of each machine every week in order to keep its own machinery working. Let x be the number of type A machine made in the week and y be the number of type B. Write down four inequalities which must be satisfied for the week. Plot them on graph paper, showing the area in which x and y can lie. If the profit made on machine A is £400 and on each machine B is £800, find the maximum profit possible.

x is the number of machine A made in the week $\therefore x \geqslant 2$.
y is the number of machine B made in the week $\therefore y \geqslant 2$.

The cost of making x of $A = £3\,000x$; y of B costs £4 000y
$\therefore$ the inequality required is $3\,000x + 4\,000y \leqslant 36\,000$.
This simplifies to $3x + 4y \leqslant 36$.

x of the A machines will take $100x$ man days to make.
y of the B machines will take $400y$ man days to make.
$\therefore$ the required inequality is $100x + 400y \leqslant 2\,000$.
This simplifies to $x + 4y \leqslant 20$.

Plot the straight lines $x = 2$; $y = 2$; $3x + 4y = 36$; $x + 4y = 20$ using the method of intercepts for the last two, as shown on page 50. Figure 27 shows the lines drawn. The possible values of x and y will be the integer values of (x, y) in and on the quadrilateral $ABCD$.

The total profit (in £) on all machines sold in one week is $400x + 800y$. If this profit is £c then $400x + 800y = c$. This equation represents a series of parallel lines called the profit lines. Plot one of these profit lines on the graph by taking a suitable value of c, say £3 200, then $400x + 800y = 3\,200$, or $x + 2y = 8$.
When $x = 0$, $y = 4$ and when $y = 0$, $x = 8$.
$\therefore$ this profit line passes through the points $P(0, 4)$ and $Q(8, 0)$ and can be drawn as shown in the figure.

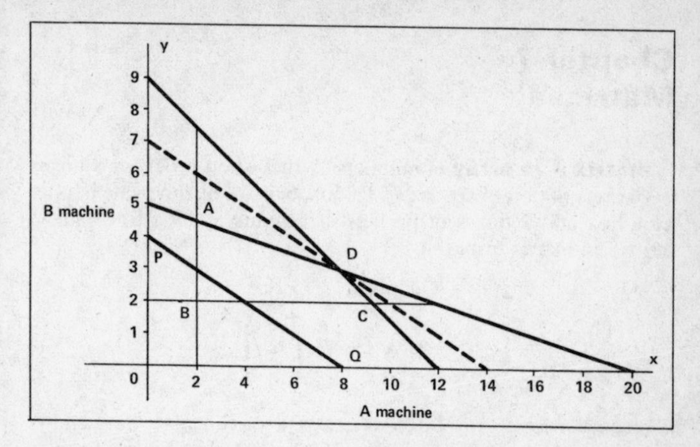

Figure 27

The line representing the maximum profit is parallel to PQ, passing through points in or on the area $ABCD$ and as far from the origin as possible, i.e. the line through D which has co-ordinates (8, 3).

$$\therefore \text{ the maximum profit} = 400 \times 8 + 800 \times 3 = 3\,200 + 2\,400$$
$$= £5\,600$$

This line is shown as a broken line on the graph.

Key terms

The symbols used are: $>$ **greater** than; $<$ **less** than; $\geqslant$ **greater** than or **equal** to; $\leqslant$ **less** than or **equal** to.
The **solution set** of an **inequation** is the range of values which satisfy the inequation.
The **locus** of a point is the path it traces, or the area it can lie in when the point is forced to move under given conditions.
Linear programming is the method of reducing certain practical problems to a number of straight lines, plotting their graphs, and using the ideas of inequalities to produce all the acceptable answers.

Chapter 7
Matrices

A **matrix** is an **array** of numbers. A tuck-shop order for 8 boxes of packets of potato crisps at £2 a box, 6 boxes of chocolate bars at £5 a box and 7 boxes of packets of peanuts at £3 a box can be written in matrix form thus:

$$\begin{matrix} \text{Quantities} & \text{Price} \\ \begin{pmatrix} 8 & 6 & 7 \end{pmatrix} & \begin{pmatrix} 2 \\ 5 \\ 3 \end{pmatrix} \end{matrix}$$

A second order is for 5 boxes of crisps, 2 of chocolate bars and 6 of peanuts and the prices have not changed. The figures for the two orders will appear as the matrices below and the cost can be worked out by multiplying the quantities by their appropriate prices:

$$\begin{pmatrix} 8 & 6 & 7 \\ 5 & 2 & 6 \end{pmatrix}\begin{pmatrix} 2 \\ 5 \\ 3 \end{pmatrix} = \begin{pmatrix} 8\times2+6\times5+7\times3 \\ 5\times2+2\times5+6\times3 \end{pmatrix} = \begin{pmatrix} 16+30+21 \\ 10+10+18 \end{pmatrix}$$

$$= \begin{pmatrix} 67 \\ 38 \end{pmatrix}$$

This is an example of matrix **multiplication**. Each element of every row of the left matrix is multiplied by the corresponding element in every column of the right matrix. Add each 'row into column'.

Example $\begin{pmatrix} 2 & 3 \\ 4 & 5 \end{pmatrix}\begin{pmatrix} a & b \\ c & d \end{pmatrix} = \begin{pmatrix} 2a+3c & 2b+3d \\ 4a+5c & 4b+5d \end{pmatrix}$

Notice the position of the elements in the answer. The top row multiplied by the right-hand column appears as the top right element of the answer.

If $A = \begin{pmatrix} 4 & 6 \\ 2 & 1 \end{pmatrix}$ $B = \begin{pmatrix} 3 & 2 \\ 5 & 1 \end{pmatrix}$ $C = (1 \quad 4)$ $D = \begin{pmatrix} 5 \\ 7 \end{pmatrix}$

$E = (6 \quad 4 \quad 2)$ $F = \begin{pmatrix} 4 & 5 \\ 8 & 3 \\ 2 & -4 \end{pmatrix}$

The order of a matrix = number of rows × number of columns.

A and B have order 2×2, F is 3×2, C and E with a single row are called row matrices and D is a column matrix. Note the following products.

$$BD \text{ or } B\times D = \begin{pmatrix} 3 & 2 \\ 5 & 1 \end{pmatrix}\begin{pmatrix} 5 \\ 7 \end{pmatrix} = \begin{pmatrix} 3\times 5+2\times 7 \\ 5\times 5+1\times 7 \end{pmatrix}$$

$$= \begin{pmatrix} 15+14 \\ 25+\ 7 \end{pmatrix} = \begin{pmatrix} 29 \\ 32 \end{pmatrix}$$

$$CD = (1 \quad 4)\begin{pmatrix} 5 \\ 7 \end{pmatrix} = (5+28) = (33)$$

$$DC = \begin{pmatrix} 5 \\ 7 \end{pmatrix}(1 \quad 4) = \begin{pmatrix} 5 & 20 \\ 7 & 28 \end{pmatrix}$$

$$EF = (6 \quad 4 \quad 2)\begin{pmatrix} 4 & 5 \\ 8 & 3 \\ 2 & -4 \end{pmatrix} = (24+32+4 \quad 30+12-8)$$

$$= (60 \quad 34)$$

Notice that the answer has the number of rows of the left-hand matrix (one) and the number of columns of the right-hand matrix (two).

$$AB = \begin{pmatrix} 4 & 6 \\ 2 & 1 \end{pmatrix}\begin{pmatrix} 3 & 2 \\ 5 & 1 \end{pmatrix} = \begin{pmatrix} 12+30 & 8+6 \\ 6+\ 5 & 4+1 \end{pmatrix} = \begin{pmatrix} 42 & 14 \\ 11 & 5 \end{pmatrix}$$

The above rule for the order of the answer holds again for AB: verify that it does for BD and CD, for it is a general rule.

Two very important points to note in matrix multiplication are:
(1) In general $AB \neq BA$. Matrix multiplication is not commutative.

$$BA = \begin{pmatrix} 3 & 2 \\ 5 & 1 \end{pmatrix}\begin{pmatrix} 4 & 6 \\ 2 & 1 \end{pmatrix} = \begin{pmatrix} 12+4 & 18+2 \\ 20+2 & 30+1 \end{pmatrix}$$

$$= \begin{pmatrix} 16 & 20 \\ 22 & 31 \end{pmatrix} \neq AB.$$

(2) Matrix multiplication is only possible when the number of columns of the **left** matrix = the number of rows of the **right** matrix.

$$FE = \begin{pmatrix} 4 & 5 \\ 8 & 3 \\ 2 & -4 \end{pmatrix} (6 \quad 4 \quad 2) = (24+5\times ? \quad 32+3\times ? \quad 4-4\times ?)$$

In attempting to multiply 'row into column' there is no corresponding element for the 5, 3 or -4, so there is no product FE. Notice that EF was evaluated above, so again $EF \neq FE$.

When AB is evaluated, B is **pre-multiplied** by A.
When BA is evaluated, B is **post-multiplied** by A.

Addition and subtraction

If a second school with a tuck-shop made two orders for crisps etc. which gave a matrix $\begin{pmatrix} 5 & 4 & 6 \\ 4 & 7 & 3 \end{pmatrix}$ then the total order for the two schools could be worked out as follows:

$$\begin{pmatrix} 8 & 6 & 7 \\ 5 & 2 & 6 \end{pmatrix} + \begin{pmatrix} 5 & 4 & 6 \\ 4 & 7 & 3 \end{pmatrix} = \begin{pmatrix} 8+5 & 6+4 & 7+6 \\ 5+4 & 2+7 & 6+3 \end{pmatrix}$$

$$= \begin{pmatrix} 13 & 10 & 13 \\ 9 & 9 & 9 \end{pmatrix}$$

Rule For addition (or subtraction) the matrices must have the same order. Add (or subtract) elements in corresponding positions.

Verify that $A+B = B+A$. Addition is commutative.

To multiply a matrix by a constant, k. $k\begin{pmatrix} a & b \\ c & d \end{pmatrix} = \begin{pmatrix} ka & kb \\ kc & kd \end{pmatrix}$

If $A = \begin{pmatrix} 4 & 6 \\ 2 & 1 \end{pmatrix}$ then

$$2A = 2 \times \begin{pmatrix} 4 & 6 \\ 2 & 1 \end{pmatrix} = \begin{pmatrix} 4 & 6 \\ 2 & 1 \end{pmatrix} + \begin{pmatrix} 4 & 6 \\ 2 & 1 \end{pmatrix} = \begin{pmatrix} 8 & 12 \\ 4 & 2 \end{pmatrix}$$

The zero matrix

If $A = \begin{pmatrix} 4 & 6 \\ 2 & 1 \end{pmatrix}$ and $O = \begin{pmatrix} 0 & 0 \\ 0 & 0 \end{pmatrix}$ then $AO = \begin{pmatrix} 0 & 0 \\ 0 & 0 \end{pmatrix}$

and $OA = \begin{pmatrix} 0 & 0 \\ 0 & 0 \end{pmatrix}$

$\therefore AO = OA = O$, an exception to the rule that $AB \neq BA$. O is called the **zero matrix**; every element has the value zero. The zero matrix can be any order. Verify also that $A+O = O+A = A$.

The unit matrix

$I = \begin{pmatrix} 1 & 0 \\ 0 & 1 \end{pmatrix}$ and $I = \begin{pmatrix} 1 & 0 & 0 \\ 0 & 1 & 0 \\ 0 & 0 & 1 \end{pmatrix}$ are examples of the **unit matrix**. It is square, with 1's in the leading diagonal and 0's in every other position.

If $A = \begin{pmatrix} 4 & 6 \\ 2 & 1 \end{pmatrix}$ then $AI = \begin{pmatrix} 4 & 6 \\ 2 & 1 \end{pmatrix}\begin{pmatrix} 1 & 0 \\ 0 & 1 \end{pmatrix}$

$$= \begin{pmatrix} 4+0 & 0+6 \\ 2+0 & 0+1 \end{pmatrix} = \begin{pmatrix} 4 & 6 \\ 2 & 1 \end{pmatrix} = A$$

Verify that $IA = A$ as well so that $AI = IA = A$, another exception to the rule that $AB \neq BA$. In algebra $a \times 1 = a$; I has the same effect in matrix multiplication so it is called the unit matrix.

If $P = \begin{pmatrix} 2 & 3 & 1 \\ 4 & 2 & 6 \end{pmatrix}$ then $PI = \begin{pmatrix} 2 & 3 & 1 \\ 4 & 2 & 6 \end{pmatrix}\begin{pmatrix} 1 & 0 & 0 \\ 0 & 1 & 0 \\ 0 & 0 & 1 \end{pmatrix}$

and $IP = \begin{pmatrix} 1 & 0 \\ 0 & 1 \end{pmatrix}\begin{pmatrix} 2 & 3 & 1 \\ 4 & 2 & 6 \end{pmatrix}$

To pre-multiply a 2×3 matrix by I, use the 2×2 form of I; to post-multiply it by I use the 3×3 form of I.

$A + O = O$ and $A \times I = A$ mean that O is the identity matrix for addition and I is the identity matrix for multiplication.

The determinant of a matrix

If $A = \begin{pmatrix} a & b \\ c & d \end{pmatrix}$ then the numerical value of $ad - bc$ is called the **determinant** of A. Only a square matrix can have a determinant. Only the 2×2 will be considered here.

If $A = \begin{pmatrix} 5 & 3 \\ 4 & -2 \end{pmatrix}$ the determinant $=$

$$5 \times -2 - 4 \times 3 = -10 - 12 = -22.$$

If the determinant is zero as in the case of $\begin{pmatrix} 6 & 8 \\ 3 & 4 \end{pmatrix}$ the matrix is called **singular**.

The inverse of a matrix

The **inverse** of a square matrix A is A^{-1}, where $A \times A^{-1} = A^{-1} \times A = I$, where I is the unit matrix. See page 23 for the analogy in algebra.

If $A = \begin{pmatrix} 4 & 5 \\ 1 & 2 \end{pmatrix}$ let $A^{-1} = \begin{pmatrix} a & b \\ c & d \end{pmatrix}$ and require

$$\begin{pmatrix} a & b \\ c & d \end{pmatrix}\begin{pmatrix} 4 & 5 \\ 1 & 2 \end{pmatrix} = \begin{pmatrix} 1 & 0 \\ 0 & 1 \end{pmatrix}$$

Multiply the matrices on the left of the equation $\begin{pmatrix} 4a+b & 5a+2b \\ 4c+d & 5c+2d \end{pmatrix} = \begin{pmatrix} 1 & 0 \\ 0 & 1 \end{pmatrix}$

Compare elements in corresponding positions on both sides:

(i) $4a + b = 1$

(ii) $5a + 2b = 0$

$8a + 2b = 2$ (i) $\times 2$

$3a = 2$ subtracting

$\Leftrightarrow a = \frac{2}{3}$

$\frac{8}{3} + b = 1$ substituting in (i)

$\Leftrightarrow b = 1 - \frac{8}{3} = -\frac{5}{3}$

(iii) $4c + d = 0$

(iv) $5c + 2d = 1$

$8c + 2d = 0$ (iii) $\times 2$

$-3c = 1$ subtracting

$\Leftrightarrow c = -\frac{1}{3}$

$-\frac{4}{3} + d = 0$ substituting in (iii)

$\Leftrightarrow d = \frac{4}{3}$

$A^{-1} \begin{pmatrix} \frac{2}{3} & -\frac{5}{3} \\ -\frac{1}{3} & \frac{4}{3} \end{pmatrix}$ is the inverse of A or $A^{-1} = \frac{1}{3}\begin{pmatrix} 2 & -5 \\ -1 & 4 \end{pmatrix}$

Compare A and A^{-1}. Carry out the following steps to find A^{-1}:

(i) find the determinant of A; $4 \times 2 - 5 \times 1 = 3$;
(ii) interchange 4 and 2, bottom right and top left elements;
(iii) 5 and 1 change signs, bottom left and top right elements;
(iv) divide each of the resulting elements by the determinant 3.

To find the inverse of $A = \begin{pmatrix} 5 & 3 \\ 2 & 1 \end{pmatrix}$

(i) the determinant $= 5 \times 1 - 3 \times 2 = 5 - 6 = -1$.
(ii) changes A to $\begin{pmatrix} 1 & 3 \\ 2 & 5 \end{pmatrix}$ (iii) makes it $\begin{pmatrix} 1 & -3 \\ -2 & 5 \end{pmatrix}$
(iv) dividing each element by -1 gives the inverse A^{-1}

$$= \begin{pmatrix} -1 & 3 \\ 2 & -5 \end{pmatrix}$$

Check: $A \times A^{-1} = \begin{pmatrix} 5 & 3 \\ 2 & 1 \end{pmatrix} \begin{pmatrix} -1 & 3 \\ 2 & -5 \end{pmatrix}$

$= \begin{pmatrix} -5+6 & 15-15 \\ -2+2 & 6-5 \end{pmatrix} = \begin{pmatrix} 1 & 0 \\ 0 & 1 \end{pmatrix}$ the unit matrix.

Verify that $A^{-1} \times A$ also gives the unit matrix.

Solving simultaneous equations using matrices

To solve $8x + 3y = 21$
$2x + y = 5$

Write them in matrix form: $\begin{pmatrix} 8 & 3 \\ 2 & 1 \end{pmatrix} \begin{pmatrix} x \\ y \end{pmatrix} = \begin{pmatrix} 21 \\ 5 \end{pmatrix}$

Note in particular how the left-hand side is written.

Pre-multiply both sides by the inverse of $\begin{pmatrix} 8 & 3 \\ 2 & 1 \end{pmatrix}$. Its determinant is $8-6=2$ $\therefore$ the inverse is $\begin{pmatrix} \frac{1}{2} & -\frac{3}{2} \\ -\frac{2}{2} & \frac{8}{2} \end{pmatrix} = \frac{1}{2}\begin{pmatrix} 1 & -3 \\ -2 & 8 \end{pmatrix}$

Pre-multiplying $\frac{1}{2}\begin{pmatrix} 1 & -3 \\ -2 & 8 \end{pmatrix}\begin{pmatrix} 8 & 3 \\ 2 & 1 \end{pmatrix}\begin{pmatrix} x \\ y \end{pmatrix}$

$= \frac{1}{2}\begin{pmatrix} 1 & -3 \\ -2 & 8 \end{pmatrix}\begin{pmatrix} 21 \\ 5 \end{pmatrix}$

Since $A^{-1} \times A = I$, $\begin{pmatrix} 1 & 0 \\ 0 & 1 \end{pmatrix}\begin{pmatrix} x \\ y \end{pmatrix}$

$= \frac{1}{2}\begin{pmatrix} 1 & -3 \\ -2 & 8 \end{pmatrix}\begin{pmatrix} 21 \\ 5 \end{pmatrix}$

Multiply the matrices $\therefore \begin{pmatrix} x \\ y \end{pmatrix} = \frac{1}{2}\begin{pmatrix} 21-15 \\ -42+40 \end{pmatrix}$

$= \frac{1}{2}\begin{pmatrix} 6 \\ -2 \end{pmatrix} = \begin{pmatrix} 3 \\ -1 \end{pmatrix}$

$\therefore x = 3 \quad y = -1$ is the solution.

Remember to pre-multiply both sides by the inverse and notice how this must reduce the left-hand side to $\begin{pmatrix} 1 & 0 \\ 0 & 1 \end{pmatrix}\begin{pmatrix} x \\ y \end{pmatrix}$ every time, because of the inverse matrix multiplication property.

The transpose of a matrix

If $A = \begin{pmatrix} 4 & 2 \\ 3 & 7 \end{pmatrix}$ is rearranged to become $A' = \begin{pmatrix} 4 & 3 \\ 2 & 7 \end{pmatrix}$ the rows and columns are interchanged. A' is called the **transpose** of A.

The number of solutions

Consider the following pairs of equations:

(i) $2x + 3y = 4$
$6x + 9y = 12$

(ii) $4x + y = 2$
$8x + 2y = 6$

(i) Solve as above: $\begin{pmatrix} 2 & 3 \\ 6 & 9 \end{pmatrix}\begin{pmatrix} x \\ y \end{pmatrix} = \begin{pmatrix} 4 \\ 12 \end{pmatrix}$

The determinant $= 18-18 = 0$. It is not possible to divide by 0 $\therefore$ there is no inverse. Observing the equations again, $6x+9y = 12$ is the other equation multiplied by 3. So there is effectively only one equation, i.e. $2x+3y = 4$. There are an infinite number of points (x, y) on the line, $\therefore$ an infinite number of solutions. $(0{\cdot}5, 1); (2, 0); (5, -2)$, etc.

(ii) $\begin{matrix} 4x + y = 2 \\ 8x + 2y = 6 \end{matrix} \Leftrightarrow \begin{pmatrix} 4 & 1 \\ 8 & 2 \end{pmatrix}\begin{pmatrix} x \\ y \end{pmatrix} = \begin{pmatrix} 2 \\ 6 \end{pmatrix}$

The determinant is $8-8 = 0$. Divide the second equation by 2 and $4x+y = 3$. But $4x+y \neq 2$ and 3 at the same time. There are no solutions. These are equations of parallel lines, which do not meet.

To determine the number of solutions:

1. Evaluate the determinant. If it is 0, there is either (a) an infinite number of solutions, or (b) no solutions. Examine the equation to find which is true.
2. Evaluate the determinant. If it is not 0, the equations will have one normal solution for x and y.

Example If $A = \begin{pmatrix} 3 & 2 \\ -1 & 4 \end{pmatrix}$ $B = \begin{pmatrix} 4 & 5 \\ 0 & 1 \end{pmatrix}$ $C = \begin{pmatrix} 6 & -3 \\ 7 & -2 \end{pmatrix}$

find the value of

(i) A^2; (ii) $A(B+C)$; (iii) $AB + AC$; (iv) $A^2 + 2I$.

(i)
$$A^2 = A \times A = \begin{pmatrix} 3 & 2 \\ -1 & 4 \end{pmatrix}\begin{pmatrix} 3 & 2 \\ -1 & 4 \end{pmatrix}$$
$$= \begin{pmatrix} 9-2 & 6+8 \\ -3-4 & -2+16 \end{pmatrix} = \begin{pmatrix} 7 & 14 \\ -7 & 14 \end{pmatrix}$$

(ii) Evaluate the bracket first $(B+C) = \begin{pmatrix} 4 & 5 \\ 0 & 1 \end{pmatrix} + \begin{pmatrix} 6 & -3 \\ 7 & -2 \end{pmatrix}$

$$= \begin{pmatrix} 10 & 2 \\ 7 & -1 \end{pmatrix}$$

$$A(B+C) = \begin{pmatrix} 3 & 2 \\ -1 & 4 \end{pmatrix} \times \begin{pmatrix} 10 & 2 \\ 7 & -1 \end{pmatrix}$$

$$= \begin{pmatrix} 30+14 & 6-2 \\ -10+28 & -2-4 \end{pmatrix} = \begin{pmatrix} 44 & 4 \\ 18 & -6 \end{pmatrix}$$

(iii) $AB = \begin{pmatrix} 3 & 2 \\ -1 & 4 \end{pmatrix}\begin{pmatrix} 4 & 5 \\ 0 & 1 \end{pmatrix} = \begin{pmatrix} 12 & 17 \\ -4 & -1 \end{pmatrix}$

$$AC = \begin{pmatrix} 3 & 2 \\ -1 & 4 \end{pmatrix}\begin{pmatrix} 6 & -3 \\ 7 & -2 \end{pmatrix} = \begin{pmatrix} 32 & -13 \\ 22 & -5 \end{pmatrix}$$

$$AB + AC = \begin{pmatrix} 12 & 17 \\ -4 & -1 \end{pmatrix} + \begin{pmatrix} 32 & -13 \\ 22 & -5 \end{pmatrix} = \begin{pmatrix} 44 & 4 \\ 18 & -6 \end{pmatrix}$$

(iv) $2I = 2\begin{pmatrix} 1 & 0 \\ 0 & 1 \end{pmatrix} = \begin{pmatrix} 2 & 0 \\ 0 & 2 \end{pmatrix}$ A^2 was evaluated in (i).

$$A^2 + 2I = \begin{pmatrix} 7 & 14 \\ -7 & 14 \end{pmatrix} + \begin{pmatrix} 2 & 0 \\ 0 & 2 \end{pmatrix} = \begin{pmatrix} 9 & 14 \\ -7 & 16 \end{pmatrix}$$

Notice that (ii) and (iii) show that $A(B+C) = AB+AC$ if the order of multiplication is strictly observed.

Example Find the values of x and k if

$$\begin{pmatrix} -1 & 2 \\ 3 & k \end{pmatrix}\begin{pmatrix} x \\ 3 \end{pmatrix} = \begin{pmatrix} 8 \\ 9 \end{pmatrix}$$

Multiply out: $-x+6 = 8 \Leftrightarrow x = -2$ Substitute below,

$$3x+3k = 9 \Leftrightarrow -6+3k = 9 \Leftrightarrow 3k = 15 \Leftrightarrow k = 5$$

With k unknown the inverse matrix method is not used here.

Example A baker sells three qualities of birthday cake at £1·00, £1·50 and £2·00 respectively. In one week he sells 4 at £1·00, 6 at £1·50 and 5 at £2·00. The following week he sells 7 at £1·00, 6 at £1·50 and 3 at £2·00. (i) Write down the number of each cake sold in the two weeks as a 2×3 matrix Q. (ii) Write down the prices as a column matrix P and evaluate QP. What does this give? (iii) Pre-multiply QP by $(1 \quad 1)$. What does this result give? (iv) Write down the matrix W which will give the total number of cakes sold each week, when combined with Q.

(i) $Q = \begin{pmatrix} 4 & 6 & 5 \\ 7 & 6 & 3 \end{pmatrix}$

(ii) $P = \begin{pmatrix} 1{\cdot}00 \\ 1{\cdot}50 \\ 2{\cdot}00 \end{pmatrix} \quad QP = \begin{pmatrix} 4 & 6 & 5 \\ 7 & 6 & 3 \end{pmatrix} \begin{pmatrix} 1{\cdot}00 \\ 1{\cdot}50 \\ 2{\cdot}00 \end{pmatrix}$

$= \begin{pmatrix} 4+9+10 \\ 7+9+6 \end{pmatrix} = \begin{pmatrix} 23 \\ 22 \end{pmatrix}$

The result gives the total takings for each week.

(iii) $(1 \quad 1)\begin{pmatrix} 23 \\ 22 \end{pmatrix} = (23+22) = (45)$. Pre-multiplying by $(1 \quad 1)$ adds the 23 and 22 $\therefore$ gives the total takings for the fortnight.

(iv) a matrix is required to produce $4+6+5$ and $7+6+3$.

Consider $QW = \begin{pmatrix} 4 & 6 & 5 \\ 7 & 6 & 3 \end{pmatrix} \begin{pmatrix} 1 \\ 1 \\ 1 \end{pmatrix}$

giving $\begin{pmatrix} 4+6+5 \\ 7+6+3 \end{pmatrix} = \begin{pmatrix} 15 \\ 16 \end{pmatrix}$ the required result.

$\therefore$ Q must be post-multiplied by $W = \begin{pmatrix} 1 \\ 1 \\ 1 \end{pmatrix}$.

Key terms

A **matrix** is an array of numbers.
The **order** of a matrix = number of rows × number of columns.
In the product AB of two matrices, B is **pre-multiplied** by A, or it is said that A is **post-multiplied** by B.
The **unit** matrix is a square matrix I, such that $A \times I = I \times A = A$. The value of the matrix A is unaltered under multiplication by I.
The **zero** or **null** matrix has all zero elements.
The **transpose** A' of matrix A is formed by interchanging the rows and columns of A.
The **determinant** of the matrix $\begin{pmatrix} a & b \\ c & d \end{pmatrix}$ is the numerical value of $ad-bc$. If this value is zero, the matrix is called **singular**.
The **inverse** A^{-1} of a matrix A is such that $A \times A^{-1} = A^{-1} \times A = I$, the unit matrix.

Chapter 8
Elementary Geometry

Angles

In figure 28(a) the angle *AOB*, less than 90°, is called an **acute** angle. The angle *AOC*, between 90° and 180°, is called **obtuse**. In figure 28(b) the other angle *AOB*, between 180° and 360°, is called **reflex**.

Angles which add up to 90° are called **complementary** angles. In figure 28(a) the angles *AOB* and *AOC* are called **supplementary**. They add up to 180°. In figure 28(c) the angles where two lines cross are equal, i.e., $w° = x°$ and $y° = z°$. Each pair is called **vertically opposite** angles.

In figure 28(d) l_1 and l_2 are parallel lines. l_3 is called a **transversal**. The angles *a* and *b* are equal, called **alternate** angles. Angles *a* and *c* are equal, called **corresponding** angles. Angles *a* and *d* add up to 180° (i.e. supplementary angles) and are called **interior** angles.

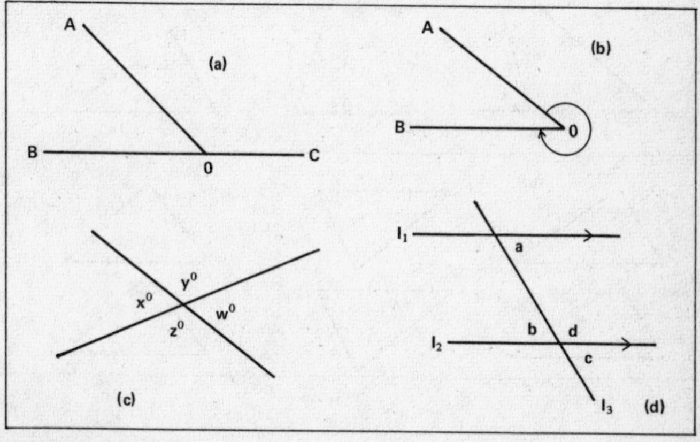

Figure 28

Symmetry

If a figure **reflected** in a line maps on to itself, the line is a line of **symmetry** and the figure possesses **bilateral** symmetry. If a figure when rotated about an axis maps on to itself, the figure possesses **rotational** symmetry. A rotation of 360° always maps a figure on to itself so this case is not worth noting.

Triangles

In figure 29(a) the sum of the angles of any triangle is 180°, i.e. $a+b+c=180°$. Note also that $c+d=180°$ (a straight line) $\therefore\ a+b=d$, i.e. **the exterior angle of a triangle equals the sum of the interior opposite angles**.

In a **scalene** triangle all the sides and angles are different. In an acute-angled triangle all the angles are less than 90°. In an obtuse-angled triangle one angle is greater than 90°, and in a right-angled triangle one angle equals 90° (notice that the other two angles are complementary).

The **isosceles** triangle in figure 29(b) has two equal sides and angles. There is one line of symmetry. In figure 29(c) the equilateral triangle has all sides equal and all angles equal to 60°. There are three lines of symmetry meeting at G. G is also the axis of rotational symmetry. The triangle maps on to itself every 120° of rotation, i.e. 3 times in one revolution. This is called **order 3 rotational** symmetry.

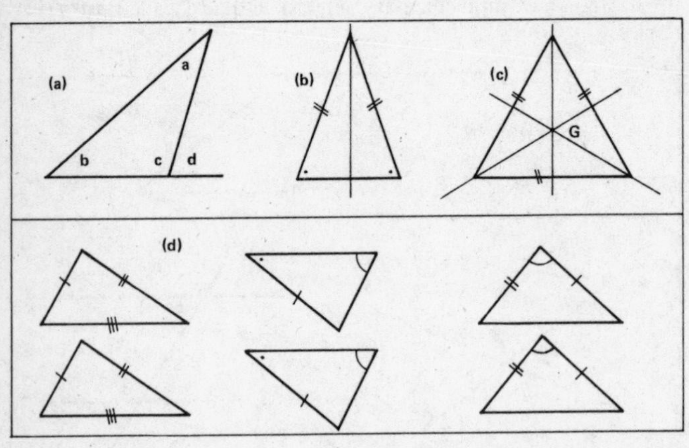

Figure 29

Figures which have the same size and shape are called **congruent**. A figure and its **image** under **rotation**, **reflection** and **translation** are congruent. Triangles are congruent if they possess one of the following sets of features: (1) the three sides of each triangle are equal; (2) two angles of each triangle are equal and a corresponding side in each triangle equal; (3) two sides in each triangle are equal and the angle between them (included angle) equal. See figure 29(d), where (d) refers to the three pairs of triangles.

Similar figures have the same shape but differ in size. Corresponding dimensions are in the same ratio. E.g. Cylinder A with radius 5 cm and height 14 cm is similar to cylinder B with radius 10 cm and height 28 cm. Both the radii and heights are in the ratio 1 : 2. B is an **enlargement** of A scale factor 2 (see page 99). Triangles will be similar if they possess one of the following sets of features: (1) the angles of each triangle are equal; (2) the corresponding sides of each triangle are in the same ratio; (3) two pairs of corresponding sides are in the same ratio and the angles between them equal.

Note that in any triangle, the sum of any two sides must be greater than the third side.

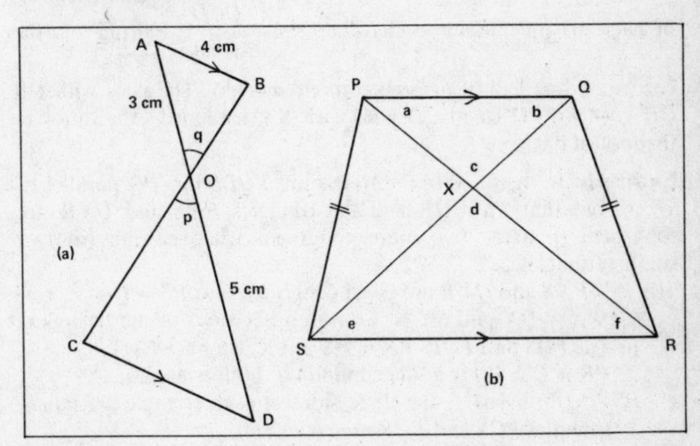

Figure 30

Example In figure 30(a) AB is parallel to CD. $AB = 4$ cm. $AO = 3$ cm. $OD = 5$ cm. $BC = 11$ cm. Calculate (i) CD; (ii) OB.

Prove the triangles similar: angle D = angle A; angle C = angle B (alternate angles AB parallel to CD). $p = q$ (vertically opposite). $\triangle AOB$ is similar to $\triangle OCD$ (equal angles).
To establish the ratios: OC corresponds to OB (opposite equal angles D and A). OD corresponds to OA (opposite angles C and B). CD corresponds to BA (opposite angles p and q).

$\frac{OC}{OB} = \frac{OD}{OA} = \frac{CD}{BA}$ are the required ratios.

(i) Let $CD = x$ cm. $OA = 3$ cm and $OD = 5$ cm. Using the ratios

$$\frac{CD}{BA} = \frac{OD}{OA} \text{ we have } \frac{x}{4} = \frac{5}{3} \Leftrightarrow x = \frac{20}{3} \quad \therefore \quad CD = 6{\cdot}67 \text{ cm.}$$

(ii) Let $OB = y$ cm then $OC = (11 - y)$ cm. From $\frac{OC}{OB} = \frac{OD}{OA}$ we have

$$\frac{11-y}{y} = \frac{5}{3} \Leftrightarrow 3(11-y) = 5y \Leftrightarrow 33 - 3y = 5y \Leftrightarrow 33 = 8y$$

$$\Leftrightarrow y = \frac{33}{8} \quad \therefore \quad OB = 4{\cdot}125 \text{ cm.}$$

The ratios can be achieved by arranging the corresponding points of each triangle under each other thus: $\frac{OCD}{OBA}$. Pairing off the corresponding letters gives corresponding sides. OC goes with OB OD goes with OA and CD goes with BA, leading to the ratios at the foot of page 83.

Example In figure 30(b) the trapezium $PQRS$ has PQ parallel to SR. Given that $PS = QR$ and the triangles PXS and QXR are congruent (i) name two more congruent triangles and (ii) two similar triangles.

(i) $\triangle$'s PXS and QXR are given congruent and $PS = QR$
$\therefore$ $PX = QX$ and $SX = XR$ (the other sides of the triangles).
In $\triangle$'s PQS and PQR, $PR = PX + XR$; $QS = QX + XS$
$\therefore$ $PR = QS$; PQ is a side common to both triangles;
$PS = QR$ (given) $\therefore$ the three sides of each triangle are equal,
$\therefore$ triangles PQS and PQR are congruent.

(ii) PQ is parallel to RS $\therefore$ angle e = angle b, and $a = f$ (alternate angles). Angle c = angle d (vertically opposite), $\therefore$ the $\triangle$'s PXQ and SXR are similar (three angles equal).

Areas and volumes of similar figures

A and B are similar rectangles. A has length = 5 cm, breadth = 2 cm; B has length = 15 cm, breadth = 6 cm.
The sides of A: sides of B = 1 : 3.
The area of $A = 5 \times 2 = 10$ cm^2. The area of $B = 15 \times 6 = 90$ cm^2.

Area of A: the area of B = 10 : 90 = 1 : 9

I.e. When the sides are enlarged three times, the area is enlarged 9 times, or 3^2 times.

C and D are similar cuboids. C has length = 6 cm, breadth = 5 cm and height = 3 cm; D has length = 12 cm, breadth = 10 cm and height = 6 cm.

The sides of C: the sides of $D = 1:2$.
Volume of $C = 6 \times 5 \times 3 = 90$ cm^3: Volume of $D = 12 \times 10 \times 6$
$= 720$ cm^3

Volume of C: volume of $D = 90 : 720 = 1 : 8$

I.e. when the sides are enlarged 2 times the volume is enlarged 8 or 2^3 times.

In general, if the sides of similar figures are in the ratio $1 : n$ the ratio of the areas $= 1 : n^2$, the ratio of the volumes $= 1 : n^3$.

Pythagoras' theorem

In a right-angled triangle, the square on the hypotenuse is equal to the sum of the squares on the other two sides. In figure 31(a)

$a^2 = b^2 + c^2 \Leftrightarrow (1)\ b^2 = a^2 - c^2$ and $(2)\ c^2 = a^2 - b^2$

Use (1) and (2) to find another side when the hypotenuse is known.

Example In figure 31(b) find x and in figure 31(c) find y.

x is the hypotenuse $\therefore\ x^2 = 7^2 + 9^2 \Leftrightarrow x^2 = 49 + 81 \Leftrightarrow x^2 = 130$ $\Leftrightarrow x = \sqrt{130}\ \therefore\ x = 11{\cdot}4$ cm using square-root tables.

y is not the hypotenuse. Use (1) $y^2 = 8{\cdot}2^2 - 6{\cdot}4^2 = 67{\cdot}24 - 40{\cdot}96$ $= 26{\cdot}28\ \therefore\ y = 5{\cdot}13$ cm using square and square root tables.

Be prepared for the well-known Pythagorian triangles. In figure 31(a) let $b = 4$ cm, $c = 3$ cm then $a^2 = 4^2 + 3^2 = 16 + 9 = 25\ \therefore\ a = 5$ cm. The triangle with sides in the ratio $3:4:5$ is right-angled. Hence the triangles with sides 6, 8 and 10 cm and 1·5, 2, and 2·5 cm are also right-angled.

Further, triangles with sides in the ratio $5:12:13$, $8:15:17$, $7:24:25$ are also right-angled triangles, in accordance with Pythagoras.

Polygons

A **polygon** is a plane figure with any number of sides. A polygon with 3 sides is a triangle, with 4 sides a **quadrilateral**, with 5 sides a **pentagon**, 6 sides a **hexagon** and 8 sides an **octagon**.

In a **convex** polygon each angle is less than 180°. All of the polygons under consideration in this chapter will be convex. Each angular point is called a **vertex** (plural vertices).

A **regular** polygon has all sides equal and all angles equal.

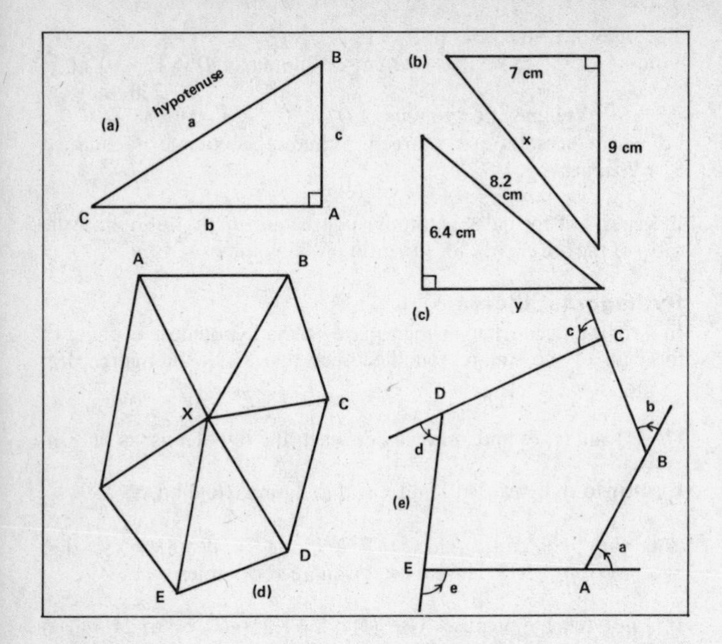

Figure 31

The sum of the interior angles of an n-sided polygon $= (2n-4) \times 90^\circ$.

To show this for $n = 6$, in figure 31(d) a point, X, is taken inside the hexagon. Each vertex is joined to X. There are now 6 triangles. The angle sum of each triangle $= 180^\circ$, $\therefore$ the sum of all the angles in the 6 triangles $= 180^\circ \times 6$.
Subtracting the 360° round X leaves the interior angles at $A, B, C \ldots F$, i.e. the sum of the interior angles $= 180^\circ \times 6 - 360^\circ$. Taking out the common factor of 90° gives $(2 \times 6 - 4)\ 90^\circ$ which is the above formula with $n = 6$. The sum is 720°.

The sum of the exterior angles of a polygon $= 360^\circ$ for all values of n (i.e. for a square, a pentagon or a hexagon).

To show this for the pentagon in figure 31(e), extend each side as shown. The exterior angle at A is a, at B is b, etc. Start at A looking along the arrow. Turn to face B, move along AB and work round the figure ending at A again. The total angle described is $a+b+c+d+e =$ one complete revolution $= 360^\circ$.

This fact is useful for dealing with regular polygons.

Example (i) Find each interior angle of a regular 12-sided polygon. (ii) How many sides has a regular polygon if each interior angle is 160°? (iii) The sum of the interior angles of a polygon is 1440°. How many sides has it?

(i) In figure 31(e) it can be seen that at each vertex:
the interior angle + the exterior angle = 180° (a straight line);
the sum of the exterior angles = 360°.
$\therefore$ each exterior angle = $360° \div 12 = 30°$.
$\therefore$ each interior angle + 30° = 180°,
$\therefore$ each interior angle = 150°.

(ii) Each exterior angle = $180° - 160° = 20°$.
The number of sides = $360° \div 20° = 18$ (1 side for each vertex).

(iii) The polygon is not regular. If n is the number of sides then $(2n-4)90° = 1440°$, the sum of the interior angles.
$\Leftrightarrow 180n - 360 = 1440 \Leftrightarrow 180n = 1800 \Leftrightarrow n = 10$. There are 10 sides.

The details of the regular pentagon and hexagon are listed below.

Each exterior angle of the **regular pentagon** = $360° \div 5 = 72°$.
Each interior angle = $180° - 72° = 108°$.
In figure 32(a) the centre is O. Angle BOA = angle AOE etc.
$\therefore$ each angle at the centre = $360° \div 5 = 72°$.
$\therefore$ the regular pentagon rotates on to itself about O every 72° (rotational symmetry order 5).

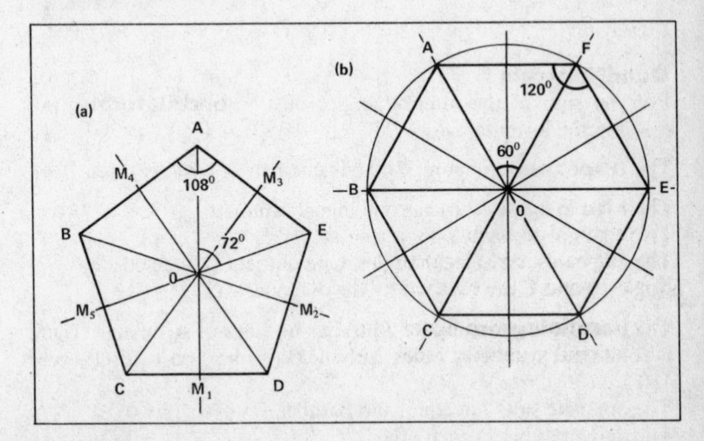

Figure 32

There are 5 lines of symmetry: $AM_1, BM_2, CM_3, DM_4, EM_5$, where M_1, M_2 etc. are the midpoints of the sides.

Each exterior angle of the **regular hexagon** $= 360° \div 6 = 60°$. Each interior angle $= 180° - 60° = 120°$. See figure 32(b).
The angles at the centre O, $AOB = BOC = \ldots = FOA = 360 \div 6 = 60°$. $\therefore$ the hexagon rotates on to itself every $60°$ (rotational symmetry order 6).
There are 6 lines of symmetry. AD, BE, CF, and the three lines joining the midpoints of opposite sides.

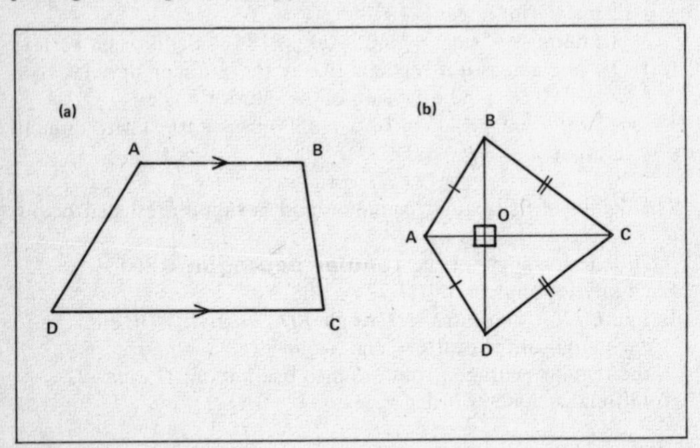

Figure 33

Quadrilaterals
For the sum of the interior angles of all **quadrilaterals**: put $n = 4$ in the formula $(2n-4)90° = (8-4)90° = 4 \times 90° = 360°$.

The **trapezium** in figure 33(a) has one pair of parallel sides.

The **kite** in figure 33(b) has one line of symmetry, AC.
Two pairs of adjacent sides are equal.
The diagonals are at right angles. One of them is bisected.
Angles A and C are bisected by the diagonal AC.

The **parallelogram** figure 34(a) has no lines of symmetry. There is rotational symmetry order 2 about O (it maps on to itself every $180°$).
The opposite sides are equal and parallel.
The diagonals bisect each other.
The opposite angles are equal.

The **rhombus** figure 34(b) has 2 lines of symmetry, *AC* and *BD*. There is rotational symmetry order 2. It is a parallelogram in which:
All 4 sides are equal. The opposite sides are parallel.
The diagonals bisect each other at right angles.
The diagonals bisect the angles of the rhombus.

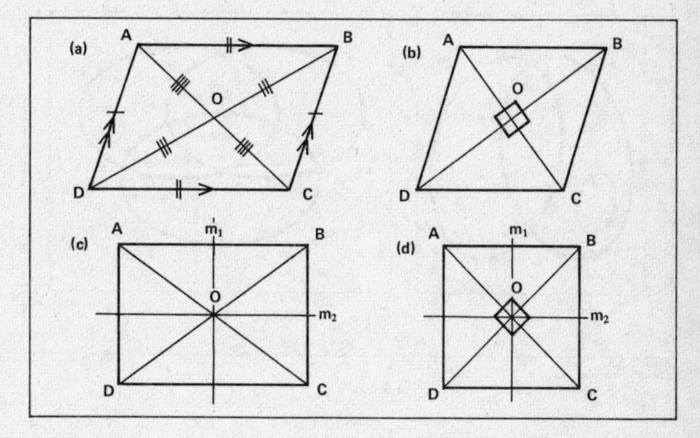

Figure 34

The **rectangle** figure 34(c) has two lines of symmetry, m_1 and m_2. There is rotational symmetry order 2. It is a parallelogram in which:
All interior angles are right angles.
The opposite sides are equal and parallel.
The diagonals bisect each other.
The diagonals are equal.

The **square** figure 34(d) has 4 lines of symmetry, *AC*, *BD*, m_1 and m_2. There is rotational symmetry order 4 (maps on to itself every 90°). It is a parallelogram in which:
All interior angles are right angles.
All 4 sides are equal. Opposite sides are parallel.
The diagonals bisect each other at right angles.
The diagonals are equal.
The diagonals bisect the angles of the square making angles of 45°.

Example Draw a Venn diagram to represent the following: $\mathscr{E} = \{\text{triangles}\}$, $A = \{\text{isosceles triangles}\}$, $B = \{\text{equilateral triangles}\}$, $C = \{\text{right-angled triangles}\}$. Identify the following: (i) $A \cap C$; (ii) $A \cap B$; (iii) $A \cap B \cap C$; (iv) $(A \cup B \cup C)'$.

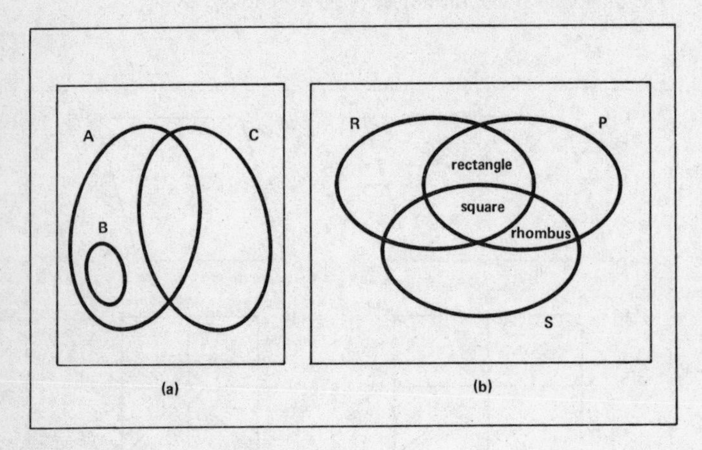

Figure 35

Draw the Venn diagram as in figure 35(a).

(i) $A \cap C = 45°, 45°, 90°$ triangles. In an isosceles triangle with one right angle, the other two angles are each 45°.

(ii) $A \cap B = B$. All equilateral triangles are isosceles.

(iii) $A \cap B \cap C = \varnothing$. All angles of an equilateral triangle are 60°. None can be 90° $\therefore$ the set is empty.

(iv) $(A \cup B \cup C)' =$ scalene triangles. No sides or angles are equal.

Example $\mathscr{E} = \{\text{quadrilaterals}\}$, $P = \{\text{parallelograms}\}$, $R = \{$all quadrilaterals with at least one right angle$\}$, $S = \{$all quadrilaterals with at least one pair of adjacent angles equal$\}$. Name (i) $R \cap P$; (ii) $S \cap P$; (iii) $R \cap P \cap S$.

Draw the Venn diagram in figure 35(b). Place the names in the spaces. A parallelogram with a right angle is a rectangle. With a pair of adjacent sides equal as well it is a square. The rhombus has no right angle but the adjacent sides are equal.

(i) $R \cap P =$ rectangles (including squares); (ii) $S \cap P =$ rhombuses (including squares); (iii) $R \cap P \cap S =$ squares only.

Circles

A chord is a line cutting a circle in two places. See figure 36(a). It divides the area into a major and a minor segment and the circumference into a major and a minor arc.

(1) The perpendicular bisector of a chord passes through the centre of the circle. In figure 36(b) notice the properties of reflection.

(2) The radius and tangent at the point of contact are at right angles. This is a special case of (1). See figure 36(c).

(3) Two tangents can be drawn from a point to a circle. They are equal in length; the line from the point to the centre bisects the angle between the tangents. OT is the axis of symmetry. $AT = BT$ and angle ATO = angle BTO. See figure 36(d).

(4) If two circles touch (a) externally, (b) internally, the distance between the centres is (a) the sum of the radii and (b) the difference between the radii. In figure 36(e) the common tangent forms a right angle with each radius $\therefore$ $O_1 O_2$ is a straight line through the point of contact of the common tangent.

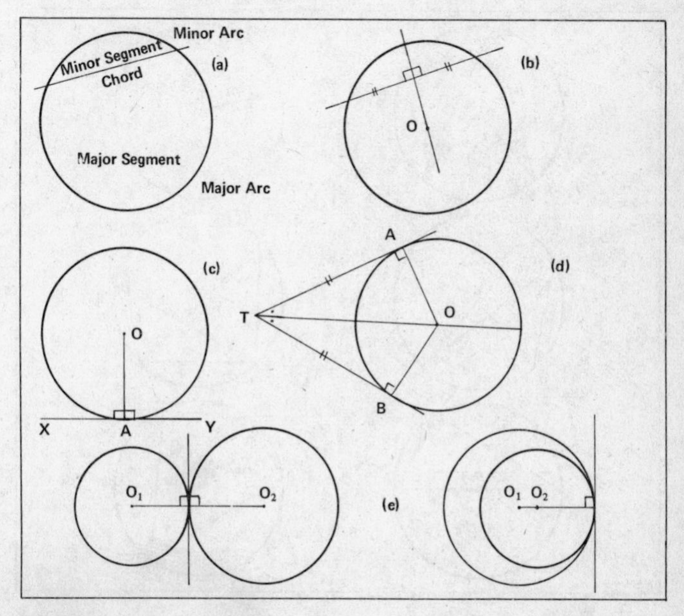

Figure 36

The angle theorems

(1) The angle which an arc subtends at the centre of a circle is twice the angle subtended by the arc at the circumference, i.e. in figure 37(a), angle AOB = twice angle ACB. Figures 37(b) and (c) show the other possible cases. The proof is now given for figure 37(a). Theorems (2), (3) and (4) all depend on this result.
Join CO and produce it. $AO = OB = OC$ (radii).
$\therefore \triangle AOC$ and $\triangle AOB$ are isosceles $\therefore p = q$ and $r = s$ (base angles)
$\therefore x = 2p$ and $y = 2r$ (exterior angles of a triangle)
$x + y = 2p + 2r = 2(p + r) \therefore AOB = 2\ ACB$.

(2) The angles subtended at the circumference by an arc are equal. In figure 37(d) $x = y$ because they both have the same angle at the centre of the circle.

(3) The angle in a semicircle is a right angle. In figure 37(e) the angle at the centre is 180°, a straight line, so the angle at the circumference is $180° \div 2 = 90°$.

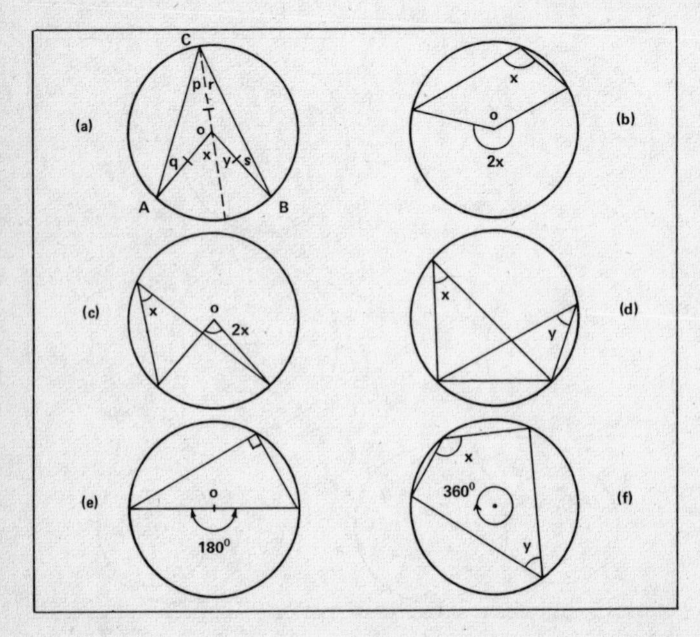

Figure 37

(4) The opposite angles of a **cyclic** quadrilateral add up to 180°. A cyclic quadrilateral has its four vertices on a circle. The four points are said to be **concyclic**. In figure 37(f) the angle at the centre is 360°, ∴ the total angle at the circumference is 180°. This total is the sum of x and y ∴ $x+y = 180°$.

Example In figure 38(a) XAY is the tangent at A, O is the centre of the circle. Angle $BAY = 48°$. Calculate (i) angle CAB; (ii) angle ACB; (iii) if $DA = DB$, find angle DAX.

Three of the above theorems are used in this example.

(i) OA is the radius, XY is the tangent at A, ∴ angle $OAY = 90°$ ∴ angle $CAB = 90° - 48° = 42°$.

(ii) AC is a diameter, ∴ angle $CBA = 90°$ (angle in semicircle). ∴ in $\triangle CAB$ angle $ACB = 180° - (90° + 42°) = 180° - 132°$
$= 48°$.

(iii) angle ADB = angle $ACB = 48°$ (subtended at the circumference by arc AB).
In $\triangle$ DAB $DA = DB$, ∴ the triangle is isosceles, and the base angles DBA and DAB are each $(180° - 48°) \div 2 = 66°$.
Angles $XAD + DAB + BAY = 180°$ (straight line).
∴ $XAD + 66° + 48° = 180°$ ∴ $XAD = 180° - 114° = 66°$.

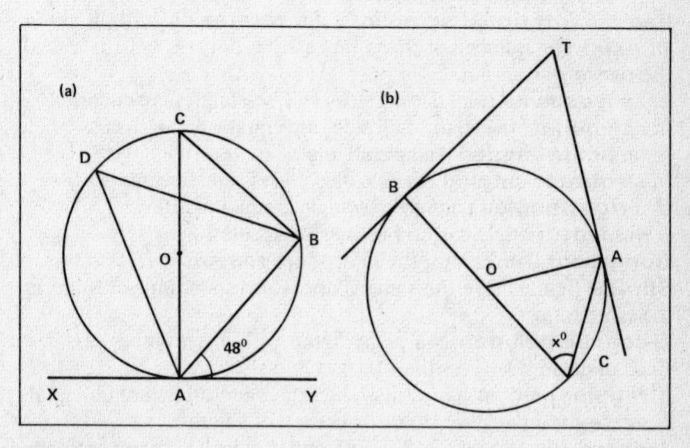

Figure 38

Notice angle BAY = angle $BDA = 48°$. This is the **alternate segment theorem**, which states that the angle between a chord and a tangent (angle BAY) is equal to the angle in the alternate segment (angle BDA).

Example In figure 38(b) AT and BT are tangents, O is the centre of the circle. Angle OCA is x°. Find in terms of x° (i) angle AOB; (ii) angle BTA; (iii) if $OCA = 4\,BTA$, find the value of x°.

(i) AOB is the angle at the centre, BCA is the angle at the circumference $\therefore AOB = 2x^\circ$.

(ii) OB and OA are radii, BT and AT are tangents $\therefore OBT = OAT = 90^\circ$. The opposite angles of the quadrilateral $AOBT$ add up to 180° $\therefore AOBT$ is a cyclic quadrilateral $\therefore AOB + BTA = 180^\circ$ $\therefore 2x^\circ + BTA = 180^\circ$, $\therefore BTA = 180^\circ - 2x^\circ$. This is in terms of x.

(iii) $OCA = 4\,BTA$ $\therefore x = 4(180^\circ - 2x) \Leftrightarrow x = 720^\circ - 8x$ $\Leftrightarrow 9x = 720^\circ \Leftrightarrow x = 80^\circ$, solving in the usual way.

Key terms

Angle A is **acute** if $A < 90^\circ$, **obtuse** if $90^\circ < A < 180^\circ$, **reflex** if $180^\circ < A < 360^\circ$.
Complementary angles add up to 90°, **supplementary** angles to 180°.
A figure has **bilateral symmetry** if it **reflects** on to itself in a **line** and **rotational symmetry** if it **rotates** on to itself about an **axis**. The number of times this can be done in 360° is called the **order** of rotation.
In an **isosceles** triangle two sides and two angles are equal.
In an **equilateral** triangle all sides are equal, all angles 60°.
In an **acute-angled** triangle all angles are less than 90°.
In an **obtuse-angled** triangle one angle is greater than 90°.
In a **right-angled** triangle one angle is equal to 90°.
A **scalene** triangle has no two measurements equal.
Congruent figures have the same shape and size.
Similar figures have the same shape. Corresponding sides are in the same ratio.
A **convex polygon** is a plane figure with any number of sides. Each of its angles is less than 180°.
A **regular** polygon has all its sides equal and all its angles equal.
A **vertex** is an angular point, or corner, of a figure.
A **chord** cuts a circle in 2 points and divides the **area** into the **major and minor segments** and the **circumference** into the **major and minor arcs**.
A **tangent touches** the circle at one point; in fact, it is a chord whose 2 points of intersection with the circle are the same.

Chapter 9
Transformation Geometry

If $\triangle ABC$ is transformed on to $\triangle A_1B_1C_1$, then $A \rightarrow A_1$, $B \rightarrow B_1$, $C \rightarrow C_1$. 'A is **mapped** on to A_1', etc. There is only one point A_1 corresponding to A: the mapping has one-to-one correspondence. Note that every point in the plane of $\triangle ABC$ maps on to a unique point on $\triangle A_1B_1C_1$. A_1 is called the **image** of A under the transformation.

An **isometry** is a transformation in which the figure and its image are the same shape and size (congruent). Reflections, rotations and translations are isometries. Shear, stretch and enlargement (see pages 99–100) are not.

Points or lines the positions of which are unaltered by a transformation are said to be **invariant**.

Reflection

In figure 39(a) line m is the **axis of reflection** (or mirror). If $OA = OA_1$ and AA_1 is perpendicular to m, then A_1 is the image of A. Conversely A is the image of A_1. If B_1 is the image of B, then line A_1B_1 is the image of AB. Similarly B_1C is the image of BC. O and C, lying on m, are invariant. $OABC$ and its image are congruent.

Table 2 shows the results of reflecting the shaded $\triangle ABC$ in figure 40 in four different axes:

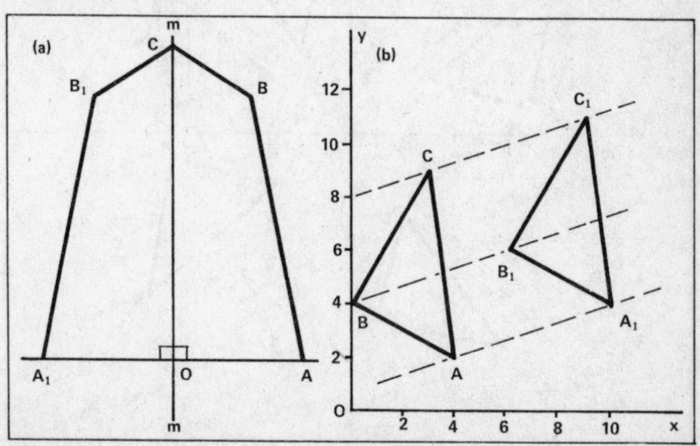

Figure 39

Axis of reflection	$A(7,1):B(8,6):C(3,2)$ map on to:		
x axis	$A_1(7,-1)$	$:B_1(8,-6)$	$:C_1(3,-2)$
y axis	$A_2(-7,1)$	$:B_2(-8,6)$	$:C_2(-3,2)$
line $y=x$	$A_3(1,7)$	$:B_3(6,8)$	$:C_3(2,3)$
line $y=-x$	$A_4(-1,-7)$	$:B_4(-6,-8)$	$:C_4(-2,-3)$

Table 2

In general a reflection in the x axis maps (x, y) on to $(x, -y)$
a reflection in the y axis maps (x, y) on to $(-x, y)$
a reflection in the line $y = x$ maps (x, y) on to (y, x)
a reflection in the line $y = -x$ maps (x, y) on to $(-y, -x)$

Translation

In figure 39(b) $\triangle ABC$ is **translated** to $\triangle A_1 B_1 C_1$ by adding 6 units to the x values and 2 to the y's. $A(4,2) \rightarrow A_1(10,4)$;

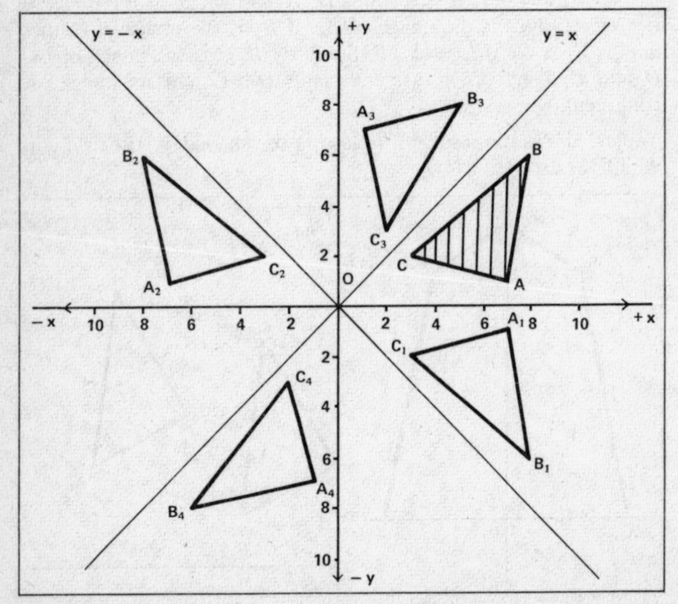

Figure 40

$B(0,4) \rightarrow B_1(6,6)$; $C(3,9) \rightarrow C_1(9,11)$. No point is invariant. The triangles are congruent. The translation is written $\begin{pmatrix} 6 \\ 2 \end{pmatrix}$. Similarly $\begin{pmatrix} -3 \\ -2 \end{pmatrix}$ moves the point (4, 5) on to (1, 3), three units to the left and two down.

Rotation

The positive direction is **anti-clockwise**. The centre of **rotation** is invariant and the figure and its image are congruent.

In figure 41(a) line l_1 is rotated about O through $+60°$. Draw OX_1 at right angles to l_1. Rotate OX_1 through $60°$ anti-clockwise. Assuming l_1 is fixed to OX_1, it too rotates through $+60°$ to position l_2.

The sum of the interior angles of a quadrilateral $= 360°$
$\therefore$ angle $a = 360° - (90° + 90° + 60°) = 360° - 240° = 120°$.
$\therefore$ angle $b = 180° - 120° = 60°$ (a and b form a straight line)
$\therefore$ the angle between l_1 and its image $l_2 = 60°$ the angle of rotation.

The following method can used to construct the centre of rotation.

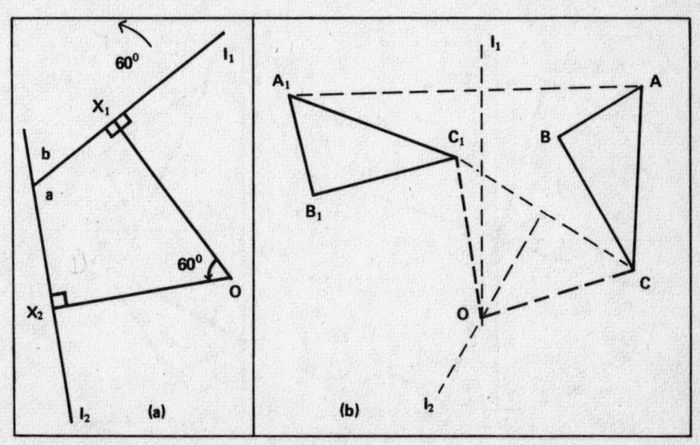

Figure 41

The positions of $\triangle ABC$ and $\triangle A_1B_1C_1$ are given in figure 41(b), but the centre and angle of rotation are unknown.

Observe that $A \to A_1$ on the circumference of a circle, the centre of which is the centre of rotation. $B \to B_1$ on a smaller circle with the same centre and $C \to C_1$ on a third circle with this centre. $\therefore AA_1$, BB_1 and CC_1 are chords of different circles with the same centre.

Using the fact that the perpendicular bisector of a chord passes through the centre of a circle: (i) Join AA_1 and CC_1; (ii) construct their perpendicular bisectors l_1 and l_2, meeting at O. The centre lies on l_1 and l_2, $\therefore O$ is the centre of rotation. Note that the perpendicular bisector of BB_1 also passes through O.

To find the angle of rotation, join OC and OC_1. Angle COC_1 is the required angle, because C describes this angle in moving to C_1. Angles AOA_1 and BOB_1 are also the angle of rotation.

In figure 42(a) the point $A(3,1)$ is rotated about the origin through (i) $+90°$ to A_1 $(-1,3)$; (ii) $180°$ (or half turn either way) to $A_2(-3,-1)$; (iii) $+270°$ to $A_3(1,-3)$.

In general a $+90°$ rotation about $(0,0)$ maps (x, y) on to $(-y, x)$
$180°$ rotation about $(0,0)$ maps (x, y) on to $(-x, -y)$
$+270°$ rotation about $(0,0)$ maps (x, y) on to $(y, -x)$

Note that a $+270°$ rotation is the same as a $-90°$ rotation.

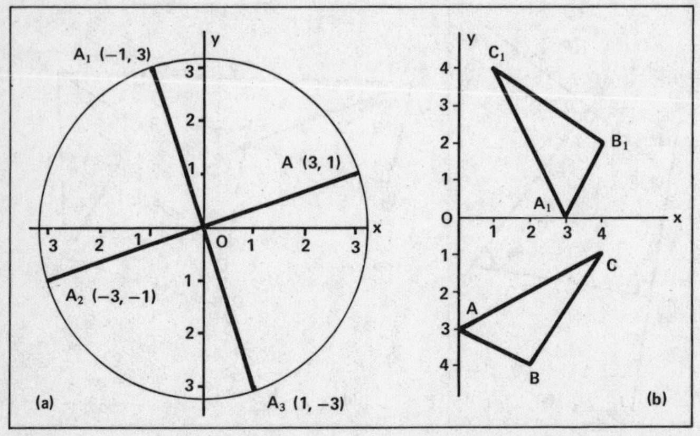

Figure 42

In figure 42(b) $A(0,-3)$ $B(2,-4)$ $C(4,-1)$ are mapped on to $A_1\,B_1\,C_1$ by a rotation of $+90^\circ$ about O. Note how A on the $-y$ axis rotates on to the $+x$ axis to the point $(3,0)$, $B \rightarrow B_1(4,2)$ and $C \rightarrow C_1(1,4)$.

Enlargement

In figure 43 $\triangle A_1\,B_1\,C_1$ is an **enlargement scale factor 2 in *O*** of $\triangle ABC$, where O is the centre of enlargement which is invariant.

$\dfrac{OA_1}{OA} = \dfrac{OB_1}{OB} = \dfrac{OC_1}{OC} = \dfrac{2}{1}$ and the triangles are similar with sides in the ratio 2 : 1. From page 84 it can be seen that the area of $\triangle A_1B_1C_1$ is 4 times that of $\triangle ABC$. The transformation is not an isometry.

$\triangle A_2B_2C_2$ is an enlargement scale factor -2 in O of $\triangle ABC$ because vectors $\overrightarrow{OA}_2$ and $\overrightarrow{OA}$ are in opposite directions $\therefore\ \dfrac{OA_2}{OA} = \dfrac{-2}{1}$.

Conversely, if $\triangle A_1B_1C_1$ is the original triangle, then $\triangle ABC$ is an enlargement of $\triangle A_1B_1C_1$ of scale factor $\frac{1}{2}$, in O.

To construct the centre of enlargement for the given $\triangle$'s ABC and $A_1B_1C_1$, join AA_1, BB_1, CC_1. They meet in O, the required centre.

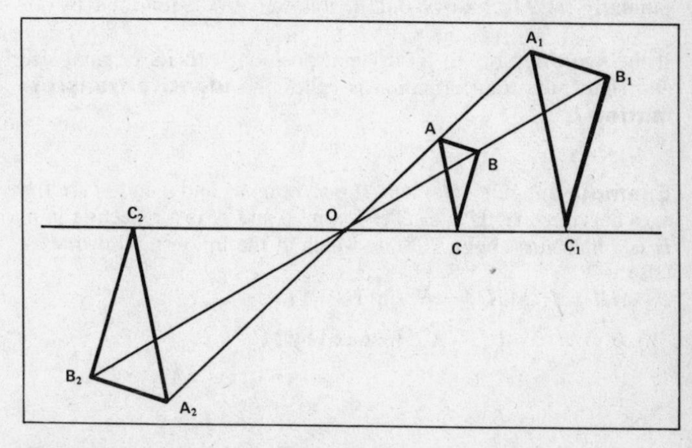

Figure 43

Shear

Figure 44(a) shows rectangle $ABCD$ transformed into parallelogram ABC_1D_1 by a **shear** parallel to the x axis in the positive direction. $\triangle DAD_1$ is removed, but congruent $\triangle CBC_1$ is added, so the area of the transformation is unaltered. The base AB is invariant, but the shape is changed, so it is not an isometry.

This figure shows that the areas of a rectangle and parallelogram on the same base and between the same parallels are equal, giving us the formula for the area of a parallelogram: base × perpendicular height.

Stretch

A figure when **stretched** is enlarged in one direction only. Figure 44(b) shows square $PQRS$ stretched into rectangle PQR_1S_1 parallel to the y axis in the positive direction and square $ABCD$ stretched into rectangle ABC_1D_1 in the positive x direction.

The product of transformations

A transformation is often given a single-letter label. If M = a specified reflection, R = a specified rotation and T = a translation, then $MR(x, y)$, or just MR, is the **product** of the **transformations**. The convention is to carry out R first on the point (x, y) and follow it by M.

Similarly for MTR carry out R, followed by T, followed by M.

If the figure returns to its original position with its original size and shape, the transformation is called the **identity transformation *I***.

Example In figure 45(a) $ABCD$ is a rhombus and n and m are the axes of symmetry. M is a reflection in m and N is a reflection in n. H is a half turn about X. State which of the following are true or false:

(i) $MH = I$; (ii) $HM = N$; (iii) $HH = I$.

(i) H gives:

$$\begin{matrix} & D & \\ A & & C \\ & B & \end{matrix}$$

followed by M:

$$\begin{matrix} & D & \\ C & & A \\ & B & \end{matrix}$$

Not the original order ∴ not I, so the statement is **false**.

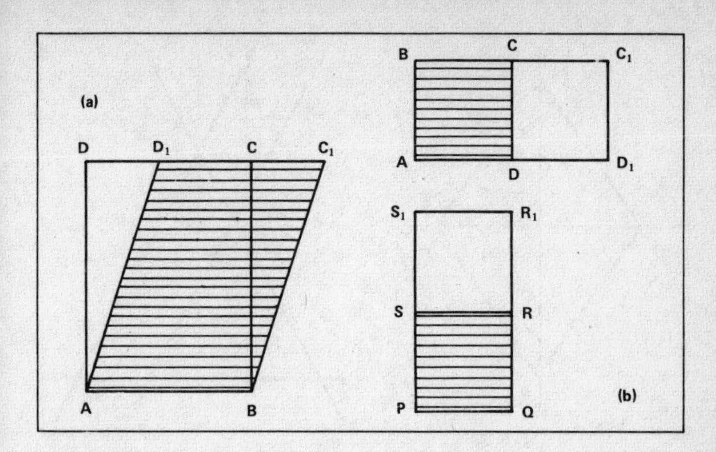

Figure 44

(ii) M gives: $\begin{matrix} & B & \\ A & & C \\ & D & \end{matrix}$ followed by H: $\begin{matrix} & D & \\ C & & A \\ & B & \end{matrix}$

This is N applied to the original, $\therefore$ the statement is **true**.

(iii) H gives: $\begin{matrix} & D & \\ A & & C \\ & B & \end{matrix}$ followed by H: $\begin{matrix} & B & \\ C & & A \\ & D & \end{matrix}$

This is the original position $\therefore$ the statement is **true**.

Verify that $NMH = I$ is also a true statement.

Example In figure 45(b): (i) find the equation of the image l_1 of l the line $y = 2x+3$ under reflection in the line $x = 1$; (ii) which point is invariant? (iii) This is followed by a translation of $\begin{pmatrix} 0 \\ -3 \end{pmatrix}$. Find the image l_2 of l_1.

(i) $y = 2x+3$ has gradient 2 and cuts the y axis at $(0, 3)$, using $y = mx+c$. It also cuts the line $x = 1$ where $y = 5$. Reflect three suitable points of l in $x = 1$.
$A(-1,1) \rightarrow A_1(3,1) : B(0,3) \rightarrow B_1(2,3)$ and $C(1,5) \rightarrow C(1,5)$.

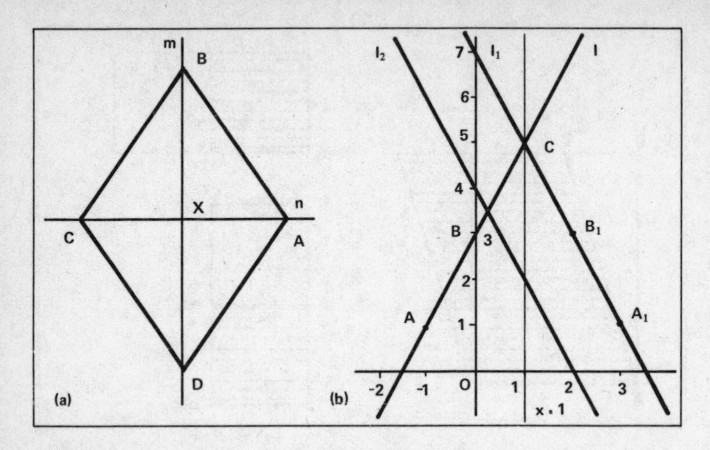

Figure 45

Joining A_1, B_1, and C we find l_1 with the same inclination to the x axis as l but with negative slope $\therefore$ its slope $= -2$.
l is of the form $y = -2x + c$ and passes through the point $(1, 5)$
$\therefore 5 = -2 + c \Leftrightarrow c = 7$. l_1 is the line $y = -2x + 7$.

(ii) $C(1, 5)$ is the invariant point.

(iii) l_2 is found by lowering l_1 3 units. $\therefore$ it cuts the y axis at $(0, 4)$. It has gradient -2, $\therefore$ l_2 is the line $y = -2x + 4$.

Example In figure 46(a) $ABCD$ is a parallelogram. $AM = \frac{1}{3}AB$: (i) describe the enlargement which maps $\triangle AMX$ on to $\triangle CDX$; (ii) find the ratio of area $\triangle AMX$ to $\triangle CDX$; (iii) name the transformation which maps $\triangle ADC$ on to $\triangle ADN$; (iv) what fraction of the parallelogram $ABCD$ is the area of the $\triangle ADN$?

(i) C is the image of A and D the image of M in X, which is the centre of the enlargement. $DC = AB$ (opposite sides of a parallelogram) $\therefore$ $DC = 3\ AM$. $\overrightarrow{XM}$ and $\overrightarrow{XD}$ are in opposite directions $\therefore$ the scale factor is -3 and the centre of enlargement X.

(ii) The sides of $\triangle CDX$ are 3 times larger than those of $\triangle AMC$ $\therefore$ the area is 9 times larger i.e. $\triangle CDX : \triangle AMX = 9:1$ (see page 84).

(iii) $\triangle ADC$ maps on to $\triangle ADN$ by a shear parallel to DA, $C \to N$.

(iv) In a shear the area of the figures remains constant,
$\therefore \triangle ADN = \triangle ADC$ in area.
$\triangle ADC = \frac{1}{2}$ parallelogram $ABCD$,
$\therefore \triangle ADN = \frac{1}{2}$ parallelogram $ABCD$.

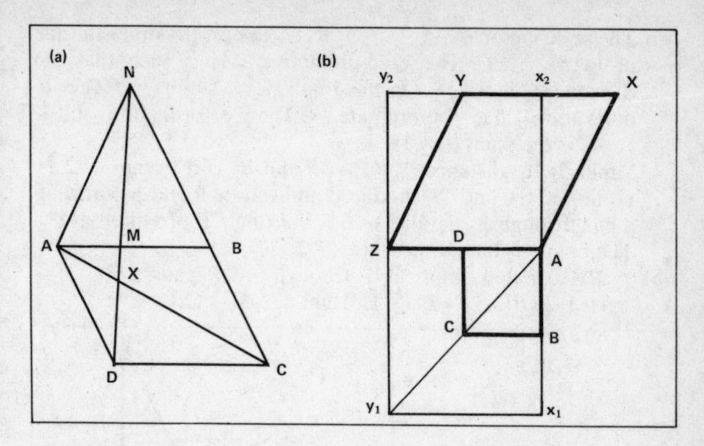

Figure 46

Example In figure 46(b) name three transformations which map the square $ABCD$ of side 2 cm into the rhombus $AXYZ$ of side 4 cm.

(1) An enlargement scale factor 2, centre A. This maps $D \to Z$, $B \to X_1$, $C \to Y_1$.

(2) Follow this by a reflection in ZA. ZA is invariant, $Y_1 \to Y_2$ $X_1 \to X_2$.

(3) Follow this by a shear parallel to the line ZA, $Y_2 \to Y$, $X_2 \to X$.

Note that the operations are commutative, i.e. can be carried out in any order.

Also note in the case of the enlargement: area $ABCD = 2 \times 2 = 4$ cm^2. The area $AX_2Y_2Z = 4 \times 4 = 16$ cm^2 so the area of the rhombus $= 16$ cm^2.

Example On graph paper plot the points $A(2,1)$, $B(3,1)$, $C(3,3)$ and $A_1(4,1)$, $B_1(7,1)$, $C_1(7,7)$. (i) Construct the centre of enlargement which maps $\triangle ABC$ on to $\triangle A_1B_1C_1$. (ii) Find the co-ordinates of the points $A_2B_2C_2$ which make $\triangle A_2B_2C_2$ an enlargement of $\triangle ABC$ of scale factor $-1{\cdot}5$, in centre P. (iii) $\triangle ABC$ is now rotated about the point $(0,0)$ through $+90°$. Find the co-ordinates of $\triangle A_3B_3C_3$ the image of $\triangle ABC$.

(i) Join C_1C and A_1A. Produce the lines to meet at $P(1,1)$. P is the centre of enlargement. $BC = 3$ units, $B_1C_1 = 6$ units, $\therefore$ the scale factor is $+2$ (+ because both triangles are on the same side of P).

(ii) The scale factor is $-\frac{3}{2}$ $\therefore$ $\triangle A_2B_2C_2$ lies on the opposite side of P to $\triangle ABC$. The x co-ordinate of A_2 is such that the distance $A_2P = \frac{3}{2}AP$. On the graph $AP = 1$ unit $\therefore$ $PA_2 = 1\frac{1}{2}$ units and A_2 has x co-ordinate $-\frac{1}{2}$. The y co-ordinate is still 1, $\therefore$ A_2 is the point $(-\frac{1}{2}, 1)$.
Similarly the distance $B_2P = \frac{3}{2}BP$ and B_2 is the point $(-2, 1)$. C_2 lies on the line CP produced and can be found by drawing a line through B_2 parallel to BC, meeting CP produced at C_2. The point C_2 has co-ordinates $(-2, -2)$.

(iii) $\triangle ABC$ rotated about $(0, 0)$ through $+90°$ causes $A(2, 1) \rightarrow A_3(-1, 2)$, $B(3, 1) \rightarrow B_3(-1, 3)$ and $C(3, 3) \rightarrow C_3(-3, 3)$.

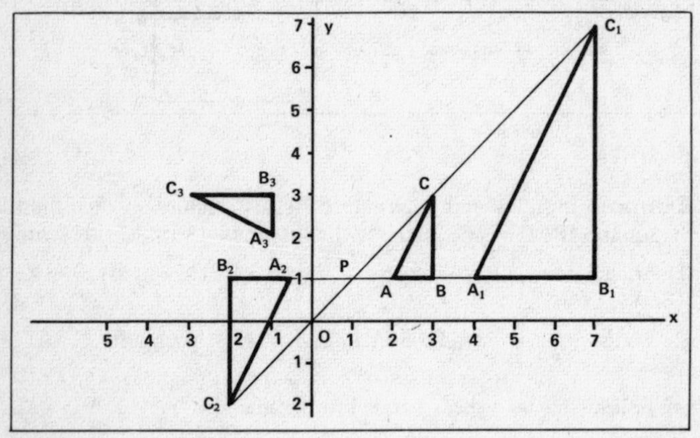

Figure 47

Key terms

In any **transformation** a point A **maps** on to its **image** A'. In an **isometry** the shape and size are unaltered. **Invariant** points and lines do not move.
A **reflection** occurs in an axis.
A **rotation** about an axis has the positive direction anti-clockwise.
In a **translation** all points move along parallel lines.
An **enlargement** in a point is defined by a **scale factor**.
In a **shear**, one line is invariant, all other points are displaced sideways. The area is unaltered.
A **stretch** alters the shape of a figure in one direction only.
The **product of transformations** MR requires R followed by M. The convention is to work from **right to left**.
The **identity** transformation returns a figure to its original position with the same shape and size.

Chapter 10
Transformation Matrices

A 2×2 **matrix** can be used to **transform** a point (x, y) in the following way: write the point as a column vector $\begin{pmatrix} x \\ y \end{pmatrix}$ and pre-multiply by the 2×2 matrix.

Reflection and rotation

Reflection in the ***x* axis**: use the matrix $\begin{pmatrix} 1 & 0 \\ 0 & -1 \end{pmatrix}$.

See figure 48(a).

We require $(2, 1) \to (2, -1)$.

$$\begin{pmatrix} 1 & 0 \\ 0 & -1 \end{pmatrix}\begin{pmatrix} 2 \\ 1 \end{pmatrix} = \begin{pmatrix} 2+0 \\ 0-1 \end{pmatrix} = \begin{pmatrix} 2 \\ -1 \end{pmatrix} \quad \text{or} \quad (2, -1)$$

Reflection in the ***y* axis**: use $\begin{pmatrix} -1 & 0 \\ 0 & 1 \end{pmatrix}$.

Write $(2, 1)$ as a column vector, pre-multiply it by $\begin{pmatrix} -1 & 0 \\ 0 & 1 \end{pmatrix}$ and show that the result is $(-2, 1)$.

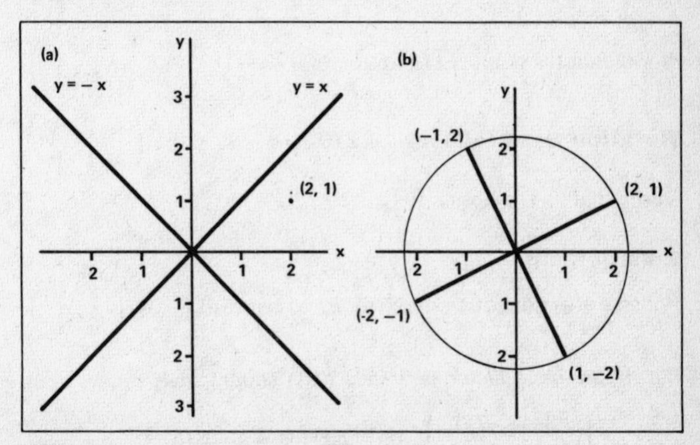

Figure 48

Reflection in the line $\boldsymbol{y=x}$: use $\begin{pmatrix} 0 & 1 \\ 1 & 0 \end{pmatrix}$. See figure 48(a).

$$\begin{pmatrix} 0 & 1 \\ 1 & 0 \end{pmatrix}\begin{pmatrix} 2 \\ 1 \end{pmatrix} = \begin{pmatrix} 0+1 \\ 2+0 \end{pmatrix} = \begin{pmatrix} 1 \\ 2 \end{pmatrix} \quad \text{or} \quad (1,2)$$

Reflection in the line $\boldsymbol{y=-x}$: use $\begin{pmatrix} 0 & -1 \\ -1 & 0 \end{pmatrix}$.

Verify that the point (2, 1) maps on to (−1, −2).

Rotation about O through **+90°**: use $\begin{pmatrix} 0 & -1 \\ 1 & 0 \end{pmatrix}$.

See figure 48(b).

We require (2, 1) to map on to (−1, 2).

$$\begin{pmatrix} 0 & -1 \\ 1 & 0 \end{pmatrix}\begin{pmatrix} 2 \\ 1 \end{pmatrix} = \begin{pmatrix} 0-1 \\ 2+0 \end{pmatrix} = \begin{pmatrix} -1 \\ 2 \end{pmatrix}$$

or (2, 1) maps on to (−1, 2).

Rotation about O through **180°** (either direction)

use: $\begin{pmatrix} -1 & 0 \\ 0 & -1 \end{pmatrix}$.

Verify that the point (2,1) maps on to (−2, −1).

Rotation about O through **+270°** (or −90°): use $\begin{pmatrix} 0 & 1 \\ -1 & 0 \end{pmatrix}$.

Verify that (2, 1) maps on to (1, −2).

Enlargement

For an **enlargement** scale factor k and centre (0, 0): use $\begin{pmatrix} k & 0 \\ 0 & k \end{pmatrix}$.

E.g. If the scale factor is 3 then (2, 1) should map on to (6, 3).

$$\begin{pmatrix} 3 & 0 \\ 0 & 3 \end{pmatrix}\begin{pmatrix} 2 \\ 1 \end{pmatrix} = \begin{pmatrix} 6 \\ 3 \end{pmatrix} \quad \text{or} \quad (2,1) \text{ maps on to } (6,3).$$

Notice that $\begin{pmatrix} -3 & 0 \\ 0 & -3 \end{pmatrix}$ maps (2, 1) on to (−6, −3). The scale factor is −3.

Translation

If the point (2, 1) is to be **translated** 3 units in the $+x$ direction and 4 in the $+y$ direction, we add 3 and 4 to the x and y co-ordinates respectively. Hence $\begin{pmatrix} 2 \\ 1 \end{pmatrix}+\begin{pmatrix} 3 \\ 4 \end{pmatrix}=\begin{pmatrix} 5 \\ 5 \end{pmatrix}$ or (2, 1) maps on to (5, 5).

Shear

In figure 49(a) $A(1,0)$ and $B(3,0)$ are on the x axis. $C(0,5)$ is to map on to the point $C_1(5,5)$ so that $\triangle ABC$ is **sheared** parallel to the x axis on to $\triangle ABC_1$.

A 2×2 matrix is required to leave the points A and B, and the y co-ordinate of C unaltered, but to change the x co-ordinate of C from 0 to 5, multiply $A(1,0)$ and $B(3,0)$ by $\begin{pmatrix} 1 & a \\ 0 & 1 \end{pmatrix}$, a being unknown.

$$\begin{pmatrix} 1 & a \\ 0 & 1 \end{pmatrix}\begin{pmatrix} 1 \\ 0 \end{pmatrix}=\begin{pmatrix} 1+0 \\ 0+0 \end{pmatrix}=\begin{pmatrix} 1 \\ 0 \end{pmatrix};\quad \begin{pmatrix} 1 & a \\ 0 & 1 \end{pmatrix}\begin{pmatrix} 3 \\ 0 \end{pmatrix}=\begin{pmatrix} 3 \\ 0 \end{pmatrix}$$

both A and B are unchanged whatever the value of a.

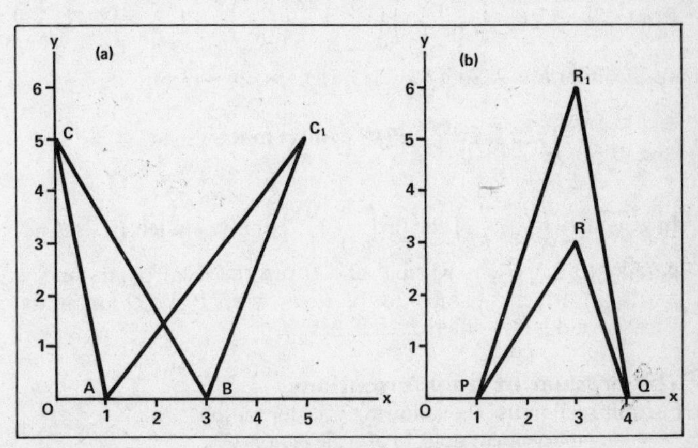

Figure 49

For $(0,5) \to (5,5)$, $\begin{pmatrix} 1 & a \\ 0 & 1 \end{pmatrix}\begin{pmatrix} 0 \\ 5 \end{pmatrix} = \begin{pmatrix} 5 \\ 5 \end{pmatrix} \Leftrightarrow \begin{pmatrix} 5a \\ 5 \end{pmatrix} = \begin{pmatrix} 5 \\ 5 \end{pmatrix}$

$\Leftrightarrow 5a = 5$ or $a = 1$. So, if $a = 1$, $(0,5)$ maps on to $(5,5)$

$\therefore$ the matrix $\begin{pmatrix} 1 & 1 \\ 0 & 1 \end{pmatrix}$ shears the triangle as required.

In general (i) $\begin{pmatrix} 1 & a \\ 0 & 1 \end{pmatrix}$ and (ii) $\begin{pmatrix} 1 & 0 \\ b & 1 \end{pmatrix}$ cause a shear by multiplication. (i) shears parallel to the x axis when the fixed points A and B are on the x axis. (ii) shears parallel to the y axis when A and B are on the y axis. a and b give different shears.

Stretch

In figure 49(b) $\triangle PQR$ is to be **stretched** on to $\triangle PQR_1$. A 2×2 matrix is required to leave P and Q unaltered, but to map $R(3,3) \to R_1(3,6)$.

Multiply $P(1,0)$ and $Q(3,0)$ by $\begin{pmatrix} 1 & 0 \\ 0 & b \end{pmatrix}$ where b is unknown.

$\begin{pmatrix} 1 & 0 \\ 0 & b \end{pmatrix}\begin{pmatrix} 1 \\ 0 \end{pmatrix} = \begin{pmatrix} 1 \\ 0 \end{pmatrix}$ $\quad \begin{pmatrix} 1 & 0 \\ 0 & b \end{pmatrix}\begin{pmatrix} 3 \\ 0 \end{pmatrix} = \begin{pmatrix} 3 \\ 0 \end{pmatrix}$

$\therefore$ P and Q are unaltered for all values of b.

For $(3,3) \to (3,6)$ $\begin{pmatrix} 1 & 0 \\ 0 & b \end{pmatrix}\begin{pmatrix} 3 \\ 3 \end{pmatrix} = \begin{pmatrix} 3 \\ 6 \end{pmatrix} \Leftrightarrow \begin{pmatrix} 3 \\ 3b \end{pmatrix} = \begin{pmatrix} 3 \\ 6 \end{pmatrix}$

$\Leftrightarrow 3b = 6$ or $b = 2$. So if $b = 2$, $(3,3)$ maps on to $(3,6)$

$\therefore \begin{pmatrix} 1 & 0 \\ 0 & 2 \end{pmatrix}$ will stretch $\triangle PQR$ parallel to the y axis.

In general (i) $\begin{pmatrix} 1 & 0 \\ 0 & b \end{pmatrix}$ and (ii) $\begin{pmatrix} a & 0 \\ 0 & 1 \end{pmatrix}$ cause a stretch. (i) stretches parallel to the y axis when P and Q are the fixed points on the x axis. (ii) stretches parallel to the x axis when P and Q are on the y axis. a and b give different stretches.

The product of transformations

For this section use the following transformations:

E = an enlargement in $(0,0)$ of scale factor 2.

R = a rotation about $(0,0)$ through $+90°$.

H = a rotation about (0, 0) through 180°.
T = a rotation about (0, 0) through +270°.
M = a reflection in the x axis. N = a reflection in the y axis.
P = a reflection in $y = x$. Q = a reflection in $y = -x$.

To obtain the **product of transformation**, multiply the matrices which give them. E.g. $EM(x, y)$ or M followed by E is in matrix form

$$\begin{pmatrix} 2 & 0 \\ 0 & 2 \end{pmatrix}\begin{pmatrix} 1 & 0 \\ 0 & -1 \end{pmatrix}\begin{pmatrix} x \\ y \end{pmatrix} = \begin{pmatrix} 2 & 0 \\ 0 & -2 \end{pmatrix}\begin{pmatrix} x \\ y \end{pmatrix} = \begin{pmatrix} 2x \\ -2y \end{pmatrix}$$

or (x, y) maps on to $(2x, -2y)$.

Pre-multiplying (x, y) by $\begin{pmatrix} 2 & 0 \\ 0 & -2 \end{pmatrix}$ reflects it in the x axis followed by an enlargement of scale factor 2 in (0, 0).

$$ME(x, y) = \begin{pmatrix} 1 & 0 \\ 0 & -1 \end{pmatrix}\begin{pmatrix} 2 & 0 \\ 0 & 2 \end{pmatrix}\begin{pmatrix} x \\ y \end{pmatrix} = \begin{pmatrix} 2 & 0 \\ 0 & -2 \end{pmatrix}\begin{pmatrix} x \\ y \end{pmatrix}$$

In this case $EM = ME$.

Now $$RM = \begin{pmatrix} 0 & -1 \\ 1 & 0 \end{pmatrix}\begin{pmatrix} 1 & 0 \\ 0 & -1 \end{pmatrix} = \begin{pmatrix} 0 & 1 \\ 1 & 0 \end{pmatrix}$$ and

$$MR = \begin{pmatrix} 1 & 0 \\ 0 & -1 \end{pmatrix}\begin{pmatrix} 0 & -1 \\ 1 & 0 \end{pmatrix} = \begin{pmatrix} 0 & -1 \\ -1 & 0 \end{pmatrix}$$

$\therefore MR \neq RM$. MR, a rotation of +90° about (0, 0) followed by a reflection in the x axis, is the same as a reflection in the line $y = -x$; and RM, a reflection in the x axis followed by a rotation of +90°, is the same as a reflection in the line $y = x$. Always carry out the transformations in the correct order.

Verify the following results using the matrices on pages 105–6.

(i) $HM = N = MH$; (ii) $NH = M = HN$; (iii) $MT = P$ but $TM = Q$.

Consider the product of three transformations:

$$MNR = \begin{pmatrix} 1 & 0 \\ 0 & -1 \end{pmatrix}\begin{pmatrix} -1 & 0 \\ 0 & 1 \end{pmatrix}\begin{pmatrix} 0 & -1 \\ 1 & 0 \end{pmatrix}$$

$$(MN)R = \begin{pmatrix} -1 & 0 \\ 0 & -1 \end{pmatrix}\begin{pmatrix} 0 & -1 \\ 1 & 0 \end{pmatrix} = \begin{pmatrix} 0 & 1 \\ -1 & 0 \end{pmatrix}$$ and

$$M(NR) = \begin{pmatrix} 1 & 0 \\ 0 & -1 \end{pmatrix}\begin{pmatrix} 0 & 1 \\ 1 & 0 \end{pmatrix} = \begin{pmatrix} 0 & 1 \\ -1 & 0 \end{pmatrix}$$

With $(MN)R$ evaluate MN first. With $M(NR)$ evaluate NR first. The results are the same and are equivalent to T. This product is associative.

The identity transformation

(i) $\begin{pmatrix} 1 & 0 \\ 0 & 1 \end{pmatrix}\begin{pmatrix} x \\ y \end{pmatrix} = \begin{pmatrix} x \\ y \end{pmatrix}$

(ii) H^2 or $HH = \begin{pmatrix} -1 & 0 \\ 0 & -1 \end{pmatrix}\begin{pmatrix} -1 & 0 \\ 0 & -1 \end{pmatrix}\begin{pmatrix} x \\ y \end{pmatrix}$

$$= \begin{pmatrix} 1 & 0 \\ 0 & 1 \end{pmatrix}\begin{pmatrix} x \\ y \end{pmatrix} = \begin{pmatrix} x \\ y \end{pmatrix}$$

(i) shows (x, y) unmoved when multiplied by the unit matrix I.

(ii) shows the result of giving (x, y) a half turn followed by a half turn. It returns to its original position and we can see that $H^2 = I$ from the working and from the geometry behind it. The **identity transformation** returns a figure to its original position, size and shape. It is equivalent to the single transformation I, the unit matrix (the identity matrix for multiplication). Verify that $(MT)P = M(TP) = I$.

Example In figure 50(a) rhombus $OABC$ is sheared so that O and B are invariant and $A(2,1) \rightarrow A_1(3,1)$. Find the transformation matrix.

Let $\begin{pmatrix} a & b \\ c & d \end{pmatrix}$ be the matrix.

Then $\begin{pmatrix} a & b \\ c & d \end{pmatrix}\begin{pmatrix} 4 \\ 0 \end{pmatrix} = \begin{pmatrix} 4 \\ 0 \end{pmatrix}$ if B is invariant.

$\Leftrightarrow \begin{pmatrix} 4a \\ 4c \end{pmatrix} = \begin{pmatrix} 4 \\ 0 \end{pmatrix} \Leftrightarrow a = 1$ and $c = 0$:

for $(2, 1) \rightarrow (3, 1)$

$$\begin{pmatrix} 1 & b \\ 0 & d \end{pmatrix}\begin{pmatrix} 2 \\ 1 \end{pmatrix} = \begin{pmatrix} 3 \\ 1 \end{pmatrix}$$

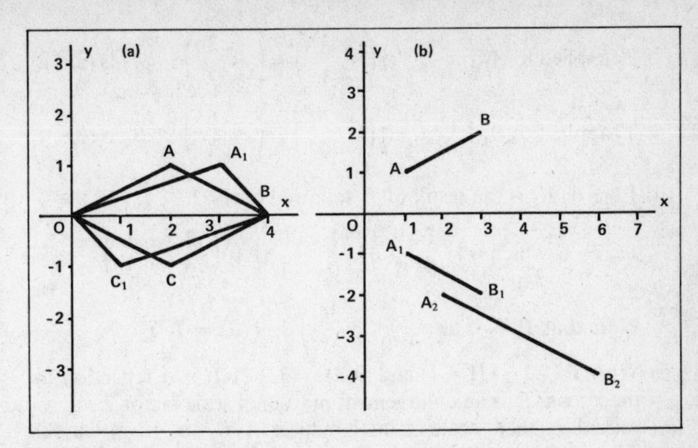

Figure 50

$$\Leftrightarrow \begin{pmatrix} 2+b \\ d \end{pmatrix} = \begin{pmatrix} 3 \\ 1 \end{pmatrix} \Leftrightarrow b = 1 \text{ and } d = 1$$

$\therefore$ the required matrix is $\begin{pmatrix} 1 & 1 \\ 0 & 1 \end{pmatrix}$.

Alternatively use the result on page 107: $\begin{pmatrix} 1 & a \\ 0 & 1 \end{pmatrix}\begin{pmatrix} 2 \\ 1 \end{pmatrix} = \begin{pmatrix} 3 \\ 1 \end{pmatrix}$ $\Leftrightarrow 2+a = 3$ or $a = 1$. This shears parallel to the x axis. Notice that $C(2,-1) \rightarrow C_1(1,-1)$.

Example The images of $A(1,1)$ and $B(3,2)$ under the transformation $T_1 = \begin{pmatrix} 1 & 0 \\ 0 & -1 \end{pmatrix}$ are A_1 and B_1. A_2 and B_2 are the images of A_1 and B_1 under $T_2 = \begin{pmatrix} 2 & 0 \\ 0 & 2 \end{pmatrix}$. (i) Find the co-ordinates of A_1, B_1, A_2 and B_2; (ii) what is the matrix of the single transformation which maps the line AB on to A_2B_2? (iii) name the transformations T_1 and T_2; (iv) find the two transformations which would map A_2B_2 on to AB. Use figure 50(b).

(i) T_1 applied to A $\begin{pmatrix} 1 & 0 \\ 0 & -1 \end{pmatrix}\begin{pmatrix} 1 \\ 1 \end{pmatrix} = \begin{pmatrix} 1 \\ -1 \end{pmatrix}$ $\therefore$ A_1 is $(1,-1)$

T_1 applied to B $\begin{pmatrix} 1 & 0 \\ 0 & -1 \end{pmatrix}\begin{pmatrix} 3 \\ 2 \end{pmatrix} = \begin{pmatrix} 3 \\ -2 \end{pmatrix}$ $\therefore$ B_1 is $(3,-2)$

T_2 applied to A_1 $\begin{pmatrix} 2 & 0 \\ 0 & 2 \end{pmatrix}\begin{pmatrix} 1 \\ -1 \end{pmatrix} = \begin{pmatrix} 2 \\ -2 \end{pmatrix}$ $\therefore$ A_2 is $(2, -2)$

T_2 applied to B_1 $\begin{pmatrix} 2 & 0 \\ 0 & 2 \end{pmatrix}\begin{pmatrix} 3 \\ -2 \end{pmatrix} = \begin{pmatrix} 6 \\ -4 \end{pmatrix}$ $\therefore$ B_2 is $(6, -4)$

(ii) Line A_2B_2 is the result of T_1 followed by T_2

i.e. T_2T_1 which is $\begin{pmatrix} 2 & 0 \\ 0 & 2 \end{pmatrix}\begin{pmatrix} 1 & 0 \\ 0 & -1 \end{pmatrix} = \begin{pmatrix} 2 & 0 \\ 0 & -2 \end{pmatrix}$

Note that T_1T_2 is also $\begin{pmatrix} 2 & 0 \\ 0 & -2 \end{pmatrix}$ $\therefore$ $T_2T_1 = T_1T_2$.

(iii) With $T_1\,(1, 1) \rightarrow (1, -1)$ and $(3, 2) \rightarrow (3, -2)$. It is a reflection in the x axis. T_2 is an enlargement in $(0, 0)$ of scale factor 2.

(iv) An enlargement scale factor $\frac{1}{2}$ reduces A_2B_2 to AB in length, followed by a reflection in the x axis. In matrix form this is $\begin{pmatrix} \frac{1}{2} & 0 \\ 0 & \frac{1}{2} \end{pmatrix}$ followed by $\begin{pmatrix} 1 & 0 \\ 0 & -1 \end{pmatrix}$ or $\begin{pmatrix} \frac{1}{2} & 0 \\ 0 & -\frac{1}{2} \end{pmatrix}$ as a single transformation.

Example $T_1: \begin{pmatrix} x \\ y \end{pmatrix} \rightarrow \begin{pmatrix} 0 & -1 \\ 1 & 0 \end{pmatrix}\begin{pmatrix} x \\ y \end{pmatrix} + \begin{pmatrix} 0 \\ -3 \end{pmatrix}$ and

$$T_2: \begin{pmatrix} x \\ y \end{pmatrix} \rightarrow \begin{pmatrix} 0 & -1 \\ 1 & 0 \end{pmatrix}\left[\begin{pmatrix} x \\ y \end{pmatrix} + \begin{pmatrix} 0 \\ -3 \end{pmatrix}\right]$$

are two transformations. Plot the points $A(2, 1)$ and $B(1, 3)$. Find (i) the images A_1B_1 under T_1 and (ii) A_2B_2 under T_2; (iii) define T_1 and T_2 in each case as one transformation followed by another.

(i) Apply T_1 to A $\begin{pmatrix} 0 & -1 \\ 1 & 0 \end{pmatrix}\begin{pmatrix} 2 \\ 1 \end{pmatrix} + \begin{pmatrix} 0 \\ -3 \end{pmatrix}$

$= \begin{pmatrix} -1 \\ 2 \end{pmatrix} + \begin{pmatrix} 0 \\ -3 \end{pmatrix} = \begin{pmatrix} -1 \\ -1 \end{pmatrix}$ $\therefore$ A_1 is $(-1, -1)$

Apply T_1 to B $\begin{pmatrix} 0 & -1 \\ 1 & 0 \end{pmatrix}\begin{pmatrix} 1 \\ 3 \end{pmatrix} + \begin{pmatrix} 0 \\ -3 \end{pmatrix}$

$= \begin{pmatrix} -3 \\ 1 \end{pmatrix} + \begin{pmatrix} 0 \\ -3 \end{pmatrix} = \begin{pmatrix} -3 \\ -2 \end{pmatrix}$ $\therefore$ B_1 is $(-3, -2)$

In T_1 multiply then add, in T_2 do the converse because of [ ]:

(ii) Apply T_2 to A $\begin{pmatrix} 0 & -1 \\ 1 & 0 \end{pmatrix}\left[\begin{pmatrix} 2 \\ 1 \end{pmatrix} + \begin{pmatrix} 0 \\ -3 \end{pmatrix}\right]$

$$= \begin{pmatrix} 0 & -1 \\ 1 & 0 \end{pmatrix} \begin{pmatrix} 2 \\ -2 \end{pmatrix} = \begin{pmatrix} 2 \\ 2 \end{pmatrix} \therefore A_2 \text{ is } (2, 2)$$

$$\text{Apply } T_2 \text{ to } B \begin{pmatrix} 0 & -1 \\ 1 & 0 \end{pmatrix} \left[\begin{pmatrix} 1 \\ 3 \end{pmatrix} + \begin{pmatrix} 0 \\ -3 \end{pmatrix} \right]$$

$$= \begin{pmatrix} 0 & -1 \\ 1 & 0 \end{pmatrix} \begin{pmatrix} 1 \\ 0 \end{pmatrix} = \begin{pmatrix} 0 \\ 1 \end{pmatrix} \therefore B_2 \text{ is } (0, 1)$$

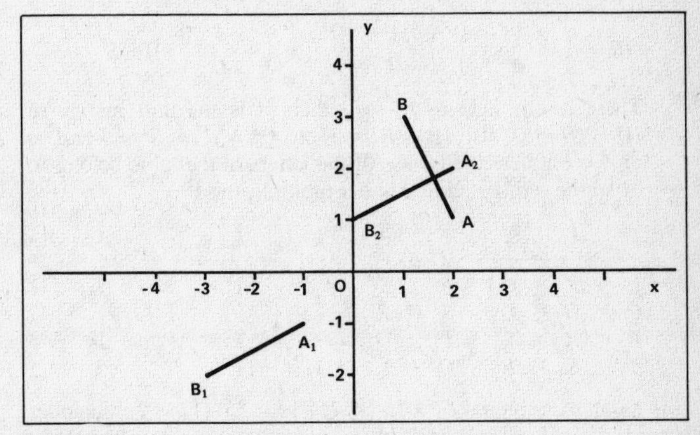

Figure 51

(iii) T_1 says carry out $\begin{pmatrix} 0 & -1 \\ 1 & 0 \end{pmatrix}$ followed by $\begin{pmatrix} 0 \\ -3 \end{pmatrix}$ (multiplication comes before addition). T_1 is a rotation about $(0, 0)$ of $+90^\circ$ followed by a translation of -3 parallel to the y axis. In T_2 the brackets [ ] signify that the translation is carried out first followed by the rotation. Notice that $T_1 T_2 \neq T_2 T_1$. These transformations can be followed from the graph in figure 51.

Example The set $N \{1, 5, 7, 11\}$ and $M = \{I, Q, R, S\}$ where

$$I = \begin{pmatrix} 1 & 0 \\ 0 & 1 \end{pmatrix} Q = \begin{pmatrix} -1 & 0 \\ 0 & -1 \end{pmatrix} R = \begin{pmatrix} 1 & 0 \\ 0 & -1 \end{pmatrix} S = \begin{pmatrix} -1 & 0 \\ 0 & 1 \end{pmatrix}$$

(i) Evaluate the transformations IQ, Q^2 and QR and by making a table show that M is closed under the binary operation of multiplication. (ii) Make a table for the binary operation $a * b$, where a and b are members of N and $*$ means 'the remainder when $a \times b$ is divided by 12'. (iii) Find the identity elements of M under multiplication and N under $*$. (iv) Find the inverse elements of M under multiplication and N under $*$. (v) The two sets and operations have the same structure. Name the corresponding elements of each set.

(i) $IQ = \begin{pmatrix} 1 & 0 \\ 0 & 1 \end{pmatrix}\begin{pmatrix} -1 & 0 \\ 0 & -1 \end{pmatrix} = \begin{pmatrix} -1 & 0 \\ 0 & -1 \end{pmatrix} = Q$

$Q^2 = \begin{pmatrix} -1 & 0 \\ 0 & -1 \end{pmatrix}\begin{pmatrix} -1 & 0 \\ 0 & -1 \end{pmatrix} = \begin{pmatrix} 1 & 0 \\ 0 & 1 \end{pmatrix} = I$

$QR = \begin{pmatrix} -1 & 0 \\ 0 & -1 \end{pmatrix}\begin{pmatrix} 1 & 0 \\ 0 & -1 \end{pmatrix} = \begin{pmatrix} -1 & 0 \\ 0 & 1 \end{pmatrix} = S$

These results help us to see others. I is the unit matrix so $IQ = QI = Q$, also $IR = RI = R$ etc. $Q^2 = R^2 = S^2 = I$ and so on to complete the table. All the letters of the table belong to M so the set M is closed under multiplication.

×	I	Q	R	S
I	I	Q	R	S
Q	Q	I	S	R
R	R	S	I	Q
S	S	R	Q	I

*	1	5	7	11
1	1	5	7	11
5	5	1	11	7
7	7	11	1	5
11	11	7	5	1

(ii) As an example take $5 * 7$; $(5 \times 7) \div 12 = 35 \div 12 = 2$ remainder $11 \therefore 5 * 7 = 11$. Work out the other pairs to complete table. All elements are members of N $\therefore$ set N is closed under $*$.

(iii) Since $IQ = QI = Q$, etc., the identity element of M is I. Since $1 * 5 = 5$, $1 * 7 = 7$, etc., the identity element of N is 1.

(iv) In M the inverse Q^{-1} of Q is such that $QQ^{-1} = I$. But $QQ = I$ $\therefore$ Q is the inverse of Q. Similarly R is the inverse of R, etc. In N the inverse of 5 is x where $5 * x = 1$. But $5 * 5 = 1$, $\therefore$ 5 is the inverse of 5. Similarly 7 is the inverse of 7, etc. Each element of M has an inverse which is a member of M. This is also true for the set N.

(v) I corresponds to 1, Q to 5, R to 7 and S to 11.

Referring to page 37, we see that (i) the two sets are closed (ii) each has an identity element in M or N and (iii) the inverse condition holds. Assuming that both are associative, we can say that M and N form groups under $\times$ and $*$ respectively. Furthermore they have the same structure, so they are isomorphic groups.

Key terms

The key terms relevant to this chapter have already been defined at the ends of chapter 7 and chapter 9.

Chapter 11
Vectors

A **scalar** quantity has **magnitude** only. A speed of 30 km/hr is a scalar. Mass and volume are other examples.

A **vector** quantity possesses both **magnitude** and **direction**. The pilot of a small aircraft must know the direction of the wind as well as its speed if he is to plot his course correctly, e.g. 30 km/hr from due south. This tells him the **velocity** of the wind. Velocity is a vector. It requires both magnitude and direction. Notice that the speed of the wind is the magnitude of the velocity. Force and displacement are other examples of vector quantities.

In figure 52(a) the line XY represents the vector $\overrightarrow{XY}$. The length XY represents the magnitude of $\overrightarrow{XY}$. The direction is specified by the arrow. Vectors are also written in the form **a**.

If $\overrightarrow{PQ}$ is such that length $PQ = XY$ and $\overrightarrow{PQ}$ is parallel to $\overrightarrow{XY}$ then the vector PQ = the vector XY, i.e. $\overrightarrow{PQ} = \overrightarrow{XY}$, or **a** = **b**.

$\overrightarrow{YX}$ refers to a vector with the same magnitude as $\overrightarrow{XY}$ but in the opposite direction. $\overrightarrow{YX} = -\overrightarrow{XY}$ or $\overrightarrow{YX} = -$**a**.

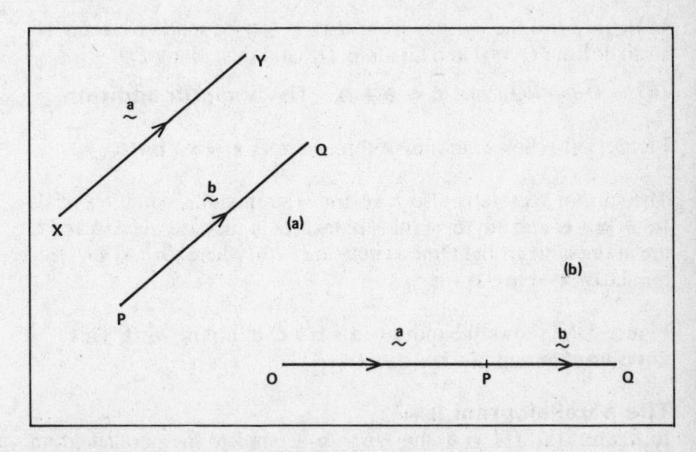

Figure 52

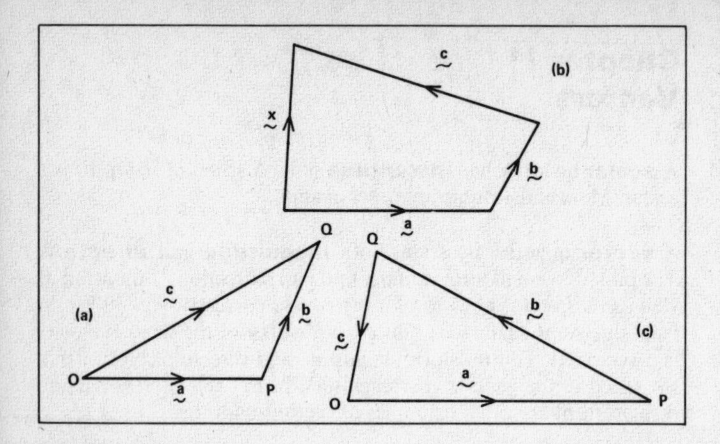

Figure 53

The **modulus** of vector $\overrightarrow{XY}$, written $|\overrightarrow{XY}|$ or $|\mathbf{a}|$, is the length of $\overrightarrow{XY}$ $\therefore$ $|\overrightarrow{XY}| = |\overrightarrow{YX}|$ or $|\mathbf{a}| = |-\mathbf{a}|$. The direction is ignored.

Addition

In figure 52(b) $\overrightarrow{OP}$ and $\overrightarrow{OQ}$ are in the same straight line.
$\overrightarrow{OQ} = \overrightarrow{OP} + \overrightarrow{PQ}$ or $\overrightarrow{OQ} = \mathbf{a} + \mathbf{b}$.

In figure 53(a) the journey from O to Q can be made by going (1) straight from O to Q and (2) along OP and then along PQ.

$\overrightarrow{OQ} = \overrightarrow{OP} + \overrightarrow{PQ}$ or $\mathbf{c} = \mathbf{a} + \mathbf{b}$. This is **vector addition**.

Figure 53(b) shows the sum of three vectors $\mathbf{x} = \mathbf{a} + \mathbf{b} + \mathbf{c}$.

The sum of vectors is also a vector. The modulus equation reads $|\mathbf{c}| = |\mathbf{a} + \mathbf{b}|$ and not $|\mathbf{c}| = |\mathbf{a}| + |\mathbf{b}|$ (except in the case where **a** and **b** are in the same straight line as in figure 52(b) where $|\overrightarrow{OQ}| = |\mathbf{a}| + |\mathbf{b}|$). Similarly $|\mathbf{x}| = |\mathbf{a} + \mathbf{b} + \mathbf{c}|$.

Figure 53(c) shows the journey $\mathbf{a} + \mathbf{b} + \mathbf{c}$, returning to O. This is a **zero vector** and $|\mathbf{a} + \mathbf{b} + \mathbf{c}| = 0$.

The parallelogram law

In figure 54(a) $\overrightarrow{OA} = \mathbf{a}$ and $\overrightarrow{OB} = \mathbf{b}$. Complete the parallelogram $OACB$. What do the diagonals OC and BA represent in figure 54(b)?

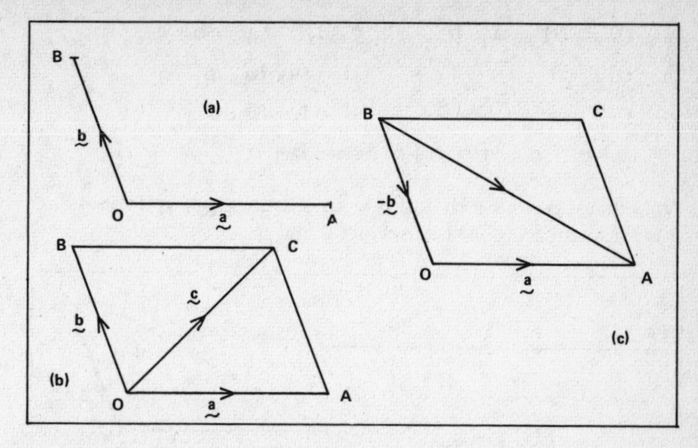

Figure 54

The opposite sides of a parallelogram are equal and parallel $\therefore \overrightarrow{BC} = \overrightarrow{OA} = \mathbf{a}$; $\overrightarrow{AC} = \overrightarrow{OB} = \mathbf{b}$,
$\therefore$ in $\triangle OAC$ **diagonal** $\overrightarrow{OC} = \mathbf{a}+\mathbf{b}$.

In $\triangle OBC$ $\overrightarrow{OC} = \overrightarrow{OB}+\overrightarrow{BC}$ or $\mathbf{c} = \mathbf{b}+\mathbf{a}$ $\therefore$ $\mathbf{b}+\mathbf{a} = \mathbf{a}+\mathbf{b}$ so **vector addition is commutative.**

In figure 54(c) consider $\triangle OAB$ to find $\overrightarrow{BA}$. $\overrightarrow{BA} = \overrightarrow{BO} + \overrightarrow{OA}$.
$\overrightarrow{OB} = \mathbf{b}$ reversing it $\overrightarrow{BO} = -\mathbf{b}$. **Diagonal** $\overrightarrow{BA} = -\mathbf{b}+\mathbf{a}$ or $\mathbf{a}-\mathbf{b}$. This is **vector subtraction**. The process of reversing a vector should be noted.

If $\overrightarrow{BA} = \mathbf{a}-\mathbf{b}$ then $\overrightarrow{AB} = \mathbf{b}-\mathbf{a}$. Thus $|\overrightarrow{AB}| = |\overrightarrow{BA}| = |\mathbf{b}-\mathbf{a}| = |\mathbf{a}-\mathbf{b}|$ directions being ignored.

Multiplication by a scalar
In figure 55(a) $\overrightarrow{OP} = \overrightarrow{PQ} = \overrightarrow{QR} = \mathbf{a}$ $\therefore$ $\overrightarrow{OR} = \mathbf{a}+\mathbf{a}+\mathbf{a} = 3\mathbf{a}$. The vector $3\mathbf{a}$ is parallel to $\mathbf{a}$ but with three times its magnitude $|3\mathbf{a}| = 3|\mathbf{a}|$.

In figure 55(b) $\overrightarrow{OX} = 2\mathbf{a}$; $\overrightarrow{XY} = 2\mathbf{b}$ $\therefore$ $\overrightarrow{OY} = 2\mathbf{a}+2\mathbf{b} = 2(\mathbf{a}+\mathbf{b})$ the distributive law holds for **scalar multiplication.**

Vector addition is also associative. In figure 55(c) $\mathbf{d} = \mathbf{a}+\mathbf{b}+\mathbf{c}$.

In $\triangle OXY$ $\overrightarrow{OY} = \mathbf{a} + \mathbf{b}$; In $\triangle XYZ$ $\overrightarrow{XZ} = \mathbf{b} + \mathbf{c}$

In $\triangle OYZ$ $\mathbf{d} = \overrightarrow{OY} + \overrightarrow{YZ}$ $\therefore$ $\mathbf{d} = (\mathbf{a}+\mathbf{b}) + \mathbf{c}$

In $\triangle OXZ$ $\mathbf{d} = \overrightarrow{OX} + \overrightarrow{XZ}$ $\therefore$ $\mathbf{d} = \mathbf{a} + (\mathbf{b}+\mathbf{c})$

$\therefore$ $(\mathbf{a}+\mathbf{b}) + \mathbf{c} = \mathbf{a} + (\mathbf{b}+\mathbf{c}) = \mathbf{a} + \mathbf{b} + \mathbf{c}$.

Note that if $\mathbf{p} = h\mathbf{a} + k\mathbf{b}$ and $\mathbf{q} = 4\mathbf{a} - 3\mathbf{b}$ then $\mathbf{p} = \mathbf{q}$ if $h = 4$ and $k = -3$, comparing the two vectors $\mathbf{p}$ and $\mathbf{q}$.

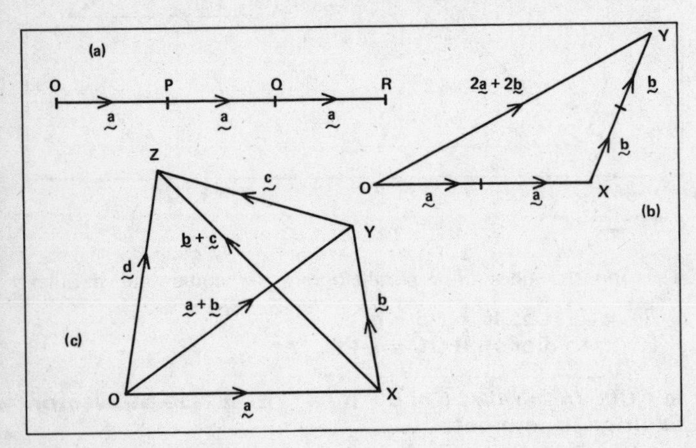

Figure 55

Position vectors

In figure 56(a) O is the origin. $\overrightarrow{OA}$ or $\mathbf{a}$ is called the **position vector** of the point A. $\overrightarrow{OB}$ or $\mathbf{b}$ is the position vector of B. If A is the point $(4, 3)$ then the vector $\overrightarrow{OA}$ is written $\begin{pmatrix} 4 \\ 3 \end{pmatrix}$ because it is equivalent to a displacement of 4 units in the direction of the x axis, followed by a displacement of 3 units in the direction of the y axis.

To find the length of $\overrightarrow{OA}$ or $|\overrightarrow{OA}|$ use Pythagoras' theorem:

$|\overrightarrow{OA}| = \sqrt{(3^2+4^2)} = \sqrt{(9+16)} = \sqrt{25}$ $\therefore$ $|\overrightarrow{OA}| = 5$ units.

B is $(1, 5)$ $\therefore$ $|\overrightarrow{OB}| = \sqrt{(1^2+5^2)} = \sqrt{(1+25)} = \sqrt{26}$

$\therefore$ $|\overrightarrow{OB}| = 5{\cdot}10$ units.

We now find the position vector of C, the fourth point of the parallelogram $OACB$ in figure 56(a). C is reached by making a displacement from $A(4,3)$ of $+1$ and $+5$ in the x and y directions respectively i.e. a move of $\begin{pmatrix}1\\5\end{pmatrix}$ $\therefore$ $\overrightarrow{OC} = \begin{pmatrix}4\\3\end{pmatrix} + \begin{pmatrix}1\\5\end{pmatrix} = \begin{pmatrix}5\\8\end{pmatrix}$

$\therefore$ C is (5,8). But $\overrightarrow{OA} = \begin{pmatrix}4\\3\end{pmatrix}$ $\overrightarrow{OB} = \begin{pmatrix}1\\5\end{pmatrix}$ $\therefore$ $\overrightarrow{OC} = \overrightarrow{OA} + \overrightarrow{OB}$ or $\mathbf{c} = \mathbf{a} + \mathbf{b}$, as is the parallelogram law.

From the same law $\overrightarrow{BA} = \mathbf{a} - \mathbf{b}$ $\overrightarrow{BA} = \begin{pmatrix}4\\3\end{pmatrix} - \begin{pmatrix}1\\5\end{pmatrix} = \begin{pmatrix}3\\-2\end{pmatrix}$.

From the diagram it can be seen that the displacement from B to A is $+3$ and -2 in the x and y directions respectively.

The components of a vector

If $\mathbf{c} = \mathbf{a} + \mathbf{b}$, $\mathbf{a}$ and $\mathbf{b}$ are called the **components** of $\mathbf{c}$. It is often convenient to take components at right angles. In figure 56(b) $P(3,2)$ has position vector $\begin{pmatrix}3\\2\end{pmatrix}$. Let $\mathbf{i}$ be a vector of modulus 1 unit along the x axis. This is called a **unit vector**. Let $\mathbf{j}$ be a unit vector along the y axis. The component of P along the x axis is $3\mathbf{i}$ and the component along the y axis is $2\mathbf{j}$.

$\therefore$ $\overrightarrow{OP} = 3\mathbf{i} + 2\mathbf{j}$ If Q is $(3,-1)$ then $\overrightarrow{OQ} = 3\mathbf{i} - \mathbf{j}$

The distance $|\overrightarrow{OP}| = \sqrt{(3^2+2^2)} = \sqrt{13} = 3{\cdot}61$ units.

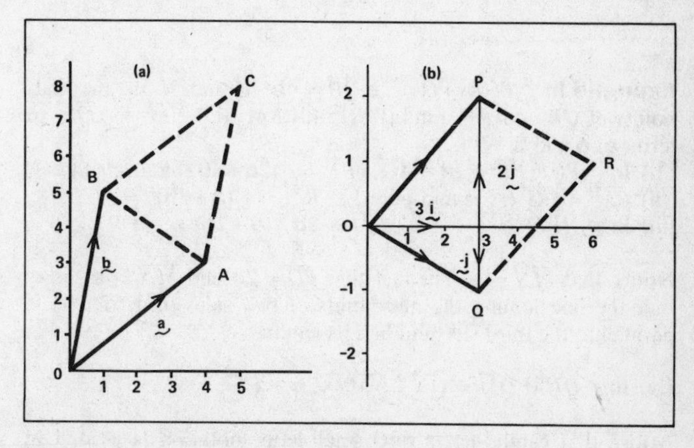

Figure 56

$\mathbf{p}+\mathbf{q} = (3\mathbf{i}+2\mathbf{j})+(3\mathbf{i}-\mathbf{j}) = 6\mathbf{i}+\mathbf{j}$.

The point $R(6,1)$ is the fourth point of the parallelogram $OPRQ$, i.e. this is an ordinary vector addition.

$\mathbf{p}-\mathbf{q} = (3\mathbf{i}+2\mathbf{j})-(3\mathbf{i}-\mathbf{j}) = 0\mathbf{i}+(2-(-1))\mathbf{j} = 3\mathbf{j}$.

The displacement from Q to P is $+3$ units along the y axis.

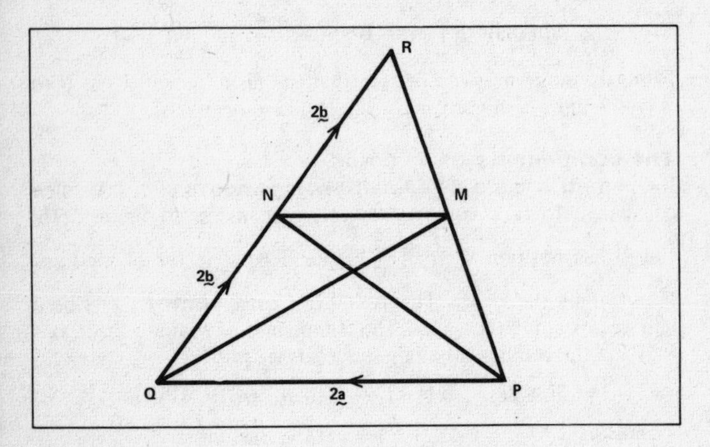

Figure 57

Example In $\triangle PQR$ $\overrightarrow{PQ} = 2\mathbf{a}, \overrightarrow{QR} = 4\mathbf{b}$, M and N are the mid-points of QR and RP. Find (i) $\overrightarrow{RP}$; (ii) $\overrightarrow{RM}$; (iii) $\overrightarrow{NM}$; (iv) $\overrightarrow{QM}$ in terms of **a** and **b**.

(i) In $\triangle PQR$ $\overrightarrow{PR} = 2\mathbf{a}+4\mathbf{b}$ $\therefore$ $\overrightarrow{RP} = -(2\mathbf{a}+4\mathbf{b})$ by reversing.
(ii) $\overrightarrow{RM} = \frac{1}{2}\overrightarrow{RP}$ (M is mid-point) $\therefore$ $\overrightarrow{RM} = -(\mathbf{a}+2\mathbf{b})$.
(iii) In $\triangle NRM$ $\overrightarrow{NM} = \overrightarrow{NR}+\overrightarrow{RM} = 2\mathbf{b}-(\mathbf{a}+2\mathbf{b}) = -\mathbf{a}$.

Notice that $\overrightarrow{MN} = \mathbf{a}$. The fact that $\overrightarrow{PQ} = 2\mathbf{a}$ and $\overrightarrow{MN} = \mathbf{a}$ shows that the line joining the mid-points of two sides of a triangle is parallel to the third side and half its length.

(iv) In $\triangle QNM$ $\overrightarrow{QM} = \overrightarrow{QN}+\overrightarrow{NM} = 2\mathbf{b}-\mathbf{a}$.

Notice that capital-letter **and** small-letter notation is used. Use whichever helps to make the vector algebra easier.

Example In figure 58(a) $\vec{OJ} = 3\mathbf{m}$, $\vec{OY} = 3\mathbf{n}$, $\vec{OC} = 12\mathbf{n}$. (i) Find $\vec{JO}$, $\vec{JY}$, $\vec{JC}$ in terms of $\mathbf{m}, \mathbf{n}$; (ii) if $\vec{SY} = \frac{1}{3}\vec{JY}$ and $\vec{JE} = \frac{1}{3}\vec{JC}$ find $\vec{OS}$, $\vec{OE}$ in terms of $\mathbf{m}, \mathbf{n}$; (iii) show that O, S and E are in the same straight line.

(i) $\vec{JO} = -\vec{OJ} = -3\mathbf{m}$

In $\triangle JOY$ $\vec{JY} = \vec{JO} + \vec{OY} = -3\mathbf{m} + 3\mathbf{n}$

$\therefore \quad \vec{JY} = 3\mathbf{n} - 3\mathbf{m}.$

In $\triangle OJC$ $\vec{JC} = \vec{JO} + \vec{OC} = -3\mathbf{m} + 12\mathbf{n}$

$\therefore \quad \vec{JC} = 12\mathbf{n} - 3\mathbf{m}.$

(ii) $\vec{SY} = \frac{1}{3}\vec{JY} = \frac{1}{3}(3\mathbf{n} - 3\mathbf{m}) \quad \therefore \quad \vec{SY} = \mathbf{n} - \mathbf{m}$

In $\triangle SOY$ $\vec{OS} = \vec{OY} + \vec{YS}$; $\vec{YS} = -\vec{SY} = \mathbf{m} - \mathbf{n}$

$\therefore \quad \vec{OS} = 3\mathbf{n} + \mathbf{m} - \mathbf{n} \quad \therefore \quad \vec{OS} = 2\mathbf{n} + \mathbf{m}.$

$\vec{JE} = \frac{1}{3}\vec{JC} = 4\mathbf{n} - \mathbf{m}.$

In $\triangle EOJ$ $\vec{OE} = \vec{OJ} + \vec{JE} = 3\mathbf{m} + 4\mathbf{n} - \mathbf{m}$

$\therefore \quad \vec{OE} = 4\mathbf{n} + 2\mathbf{m}$ or $2(2\mathbf{n} + \mathbf{m})$

(iii) $\vec{OS} = 2\mathbf{n} + \mathbf{m}$; $\vec{OE} = 2(2\mathbf{n} + \mathbf{m}) \quad \therefore \quad \vec{OE} = 2\vec{OS}$. I.e. $\vec{OE}$ and $\vec{OS}$ are parallel. They both pass through O $\therefore$ O, E, S are in the same line.

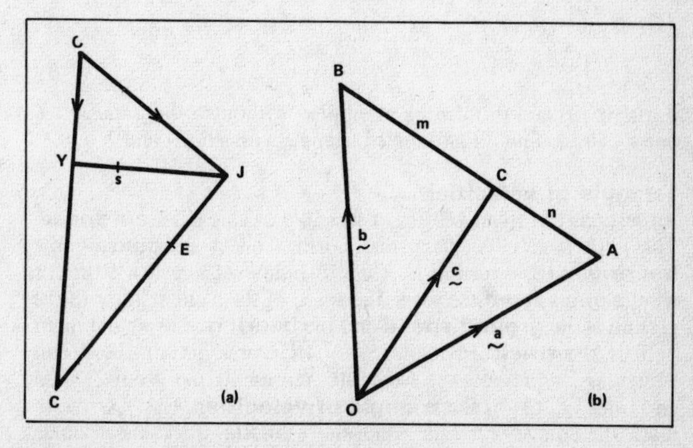

Figure 58

Example In figure 58(b) **a** and **b** are the position vectors of A and B relative to O. Find the position vector of C which divides BA in the ratio $m : n$.

Let C have position vector **c**. In (i)$\triangle OBC$ and (ii) $\triangle OAC$ (i) $\overrightarrow{BC} = -\mathbf{b}+\mathbf{c} = \mathbf{c}-\mathbf{b}$; (ii) $\overrightarrow{CA} = -\mathbf{c}+\mathbf{a} = \mathbf{a}-\mathbf{c}$.

$$\frac{\overrightarrow{BC}}{\overrightarrow{CA}} = \frac{m}{n} = \frac{\mathbf{c}-\mathbf{b}}{\mathbf{a}-\mathbf{c}} \Leftrightarrow m(\mathbf{a}-\mathbf{c}) = n(\mathbf{c}-\mathbf{b}) \Leftrightarrow$$

$$m\mathbf{a} - m\mathbf{c} = n\mathbf{c} - n\mathbf{b} \Leftrightarrow m\mathbf{a} + n\mathbf{b} = n\mathbf{c} + m\mathbf{c} \Leftrightarrow$$

$$m\mathbf{a} + n\mathbf{b} = \mathbf{c}(n+m) \Leftrightarrow \mathbf{c} = \frac{m\mathbf{a}+n\mathbf{b}}{n+m}.$$

When C is the mid-point $m = n = 1$, $\mathbf{c} = \frac{1}{2}(\mathbf{a}+\mathbf{b})$.

Example A and B have position vectors $\mathbf{a} = \begin{pmatrix} 3 \\ 4 \end{pmatrix}$, $\mathbf{b} = \begin{pmatrix} -1 \\ 2 \end{pmatrix}$.

(i) What is the vector $\mathbf{a}-3\mathbf{b}$? (ii) **i** and **j** are unit vectors along the x and y axes respectively: write **a**, **b**, $\mathbf{a}-3\mathbf{b}$ in terms of **i**, **j**. (iii) write $\mathbf{b}-\mathbf{a}$ in terms of **i**, **j**; (iv) find the length of $\overrightarrow{AB}$.

(i) $\mathbf{a} - 3\mathbf{b} = \begin{pmatrix} 3 \\ 4 \end{pmatrix} - 3\begin{pmatrix} -1 \\ 2 \end{pmatrix} = \begin{pmatrix} 3 \\ 4 \end{pmatrix} - \begin{pmatrix} -3 \\ 6 \end{pmatrix} = \begin{pmatrix} 6 \\ -2 \end{pmatrix}$

(ii) $\mathbf{a} = 3\mathbf{i} + 4\mathbf{j}$; $\mathbf{b} = -\mathbf{i} + 2\mathbf{j}$; $\mathbf{a} - 3\mathbf{b} = 6\mathbf{i} - 2\mathbf{j}$ (from (i))

(iii) $\mathbf{b} - \mathbf{a} = (-\mathbf{i}+2\mathbf{j}) - (3\mathbf{i}+4\mathbf{j}) = -4\mathbf{i} - 2\mathbf{j}$.

(iv) The length of $\overrightarrow{AB} = |\mathbf{b}-\mathbf{a}| = \sqrt{((-4)^2+(-2)^2)} = \sqrt{20}$

$\therefore\quad |\overrightarrow{AB}| = 4{\cdot}47$

Parts (i), (ii) and (iii) can be attempted without a diagram. Draw one for (iv) to find a right-angled triangle with sides 4 and 2.

Triangle of velocities

An aircraft can fly at 120 km/hr in still air. This is its **air speed**. The pilot points the plane due north. This is its **course** (the **bearing** of the direction in which it points). There is a 50 km/hr wind blowing **from** the west. The speed of the plane relative to the ground is the **ground speed** and the **bearing** of its actual flight path is the **track**. In figure 59(a) $\overrightarrow{AB}$ represents air speed and course, $\overrightarrow{BC}$ represents the wind. $\overrightarrow{AC}$ represents the ground speed and track. $\triangle ABC$ is the **triangle of velocities**. Use 1 arrow to mark the course, 2 for the wind and 3 for the track, the triangle must obey the vector addition law:

1 arrow + 2 arrow = 3 arrow. See figure 59(b).

To find the ground speed and track, in this case, use calculation. The right-angled triangle is best solved by using Pythagoras and trigonometry:

$AC = \sqrt{50^2+120^2} = \sqrt{16900} \quad \therefore \quad AC = 130.$

The ground speed is 130 km/hr. (You can spot the 5 : 12 : 13 triangle.)

To find angle BAC, $\frac{50}{120} = \tan BAC \; \therefore \; \tan BAC = 0{\cdot}4170 \; \therefore$ angle $BAC = 22°38'$. The angle between the course and track (called **drift**) is in this example the track 022°38′.

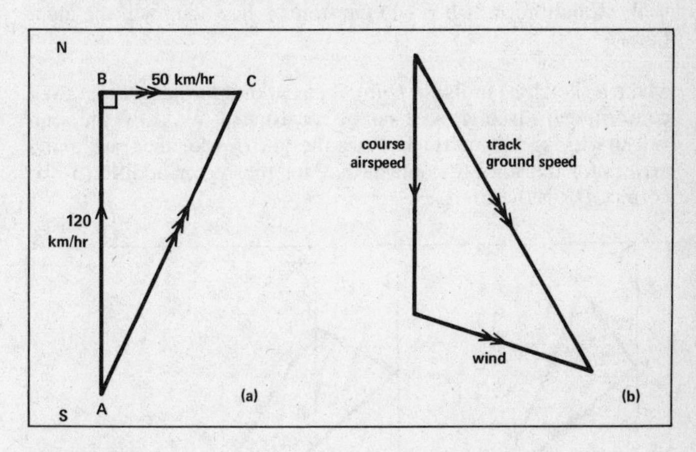

Figure 59

This diagram could also be used to solve the following problem. A man in a small boat capable of 12 km/hr in still water points the boat at right angles to the bank of a river. The current is flowing at 5 km/hr. Find the boat's speed relative to the bank and its direction taken across the river.

The speed relative to the bank replaces ground speed, the current replaces the wind, and the speed in still water replaces air speed.

As the dimensions are each 10 times smaller, the speed relative to the bank = 13 km/hr, and the angle is still 22°38′.

When the speeds and directions do not involve such convenient numbers, the option of drawing or calculation is usually given. The drawing method is advisable. Some questions also involve distances as well as speeds; take care not to confuse these on a drawing.

The following example represents the type of problem frequently confronting pilots. The pilot knows his air speed, the wind velocity and where he wants to go (track). He must find his course and the ground speed; the latter tells him how long the flight will take.

Example A plane with an air speed of 150 km/hr is to fly from A to B which is on a bearing of 130° from A. There is a 40 km/hr wind blowing from 240°. Find the course the pilot must set in order to fly straight to B. If B is 500 km from A, how long will the flight take?

Make a sketch as in figure 60(a). 3 pieces of information are given concerning the triangle so it can be constructed. We know the wind velocity, air speed and track. Place the 3 arrows for the track and 2 arrows for the wind to confirm that, for the vector addition to be correct, QR is the course.

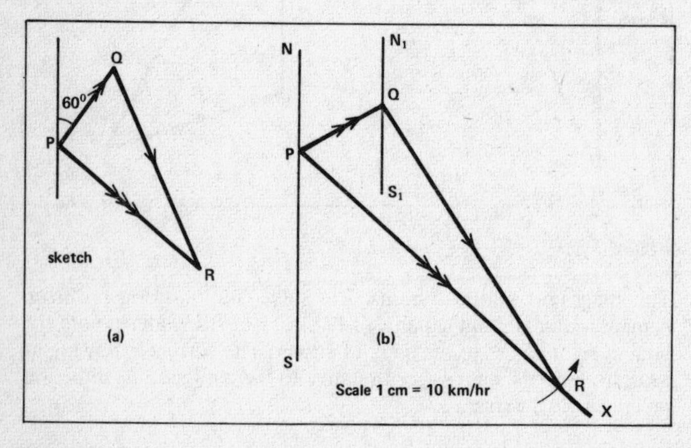

Figure 60

Choose a scale of 1 cm = 10 km/hr in figure 60(b).
Through a point P draw a north-south line. Draw PQ 4 cm long so the angle $NPQ = 60°$ (wind is from 240°).
Through P draw PX of unknown length, so that angle $NPX = 130°$.
With centre Q, radius 15 cm, draw an arc cutting PX at R.

$\triangle PQR$ is the required triangle of velocities.
Draw another north-south line through Q and angle $N_1QR =$ 144°30′ is the course to set.
The length PR is the ground speed, approximately 159 km/hr.
Now introduce the distance $AB = 500$ km for the first time. If A is taken at the point P then B is 500 km along PX. The speed in this direction is 159 km/hr and distance = speed × time.

$\therefore$ time = 500 ÷ 159 = 3·14 hrs or 3 hrs 8 minutes.

Figure 60(b) could also be used to solve the problem of a ship, capable of 15 knots in still water, about to sail from A to B, a distance of 50 nautical miles. B is on a bearing of 130° from A. The current of 4 knots is flowing from 240°. In figure 60(b) the angle N_1QR represents the course and PR the speed of the ship relative to the land. Note that 1 knot = 1 nautical mile/hr.

Representation of vectors It is important to distinguish between scalar and vector quantities. Scalars are written a, whereas in print vectors are generally shown in bold type, **a**, or, where the direction is not already specified, $\overrightarrow{AB}$. When working examples by hand, you should indicate vector quantities by underlining, $\underset{\sim}{a}$, as in the figures in this book, or simply $\underline{a}$.

Key terms

A **vector** quantity possesses magnitude and direction.
A **scalar** quantity possesses magnitude only.
The **modulus** of a vector is its positive length.
A **position vector** is used to define the distance and direction of a point A from another point O.
If **c** = **a** + **b** then **a** and **b** are called the **components** of **c**.
The **unit vectors i** and **j** are parallel to the x and y axes respectively. Each has modulus 1.
Air speed is the speed of an aircraft in still air.
Course is the bearing of the direction in which the plane is pointed.
Ground speed is the speed relative to the ground.
Track is the bearing of the direction the plane actually takes.
The wind blows **from**, a plane flies **towards** a given direction.
The **triangle of velocities** is the vector triangle formed by the three vectors such that **course + wind = track**.

Chapter 12
Basic Arithmetic

Factors

4, 3, 6 and 2 are **factors** of 12. Each one of them divides exactly into 12. $2 \times 2 \times 3$ are the **prime factors**, because they belong to the set of prime numbers.

Fractions

Despite the UK's conversion to the metric system and the corresponding increase in the use of decimals, a sound knowledge of fractions is still essential for students of mathematics. They occur in many branches, such as similar triangles, trigonometrical ratios, and in the basic processes of algebra.

The **fraction** $\frac{2}{3}$ means 2 parts out of 3. $\frac{8}{3}$ means 8 parts out of 3, i.e. 2 whole units and $\frac{2}{3}$, or $2\frac{2}{3}$. Similarly $\frac{22}{5} = 4\frac{2}{5}$. A fraction written with the numerator larger than the denominator is called an **improper** fraction.

The value of a fraction is unaltered if the numerator and the denominator are multiplied or divided by the same number.
$\frac{8}{12} = \frac{2}{3}$ dividing numerator and denominator by 4. Similarly $\frac{15}{20} = \frac{3}{4}$. This is called **cancelling**. $\frac{3}{4}$ is expressed in its **lowest terms**.

Conversely $\frac{3}{4} = \frac{15}{20}$ multiplying numerator and denominator by 5.

To evaluate $\dfrac{4\frac{1}{5}}{7} = \dfrac{4\frac{1}{5} \times 5}{7 \times 5} = \dfrac{21}{35} = \dfrac{3}{5}$ or 0·6

Multiplying numerator and denominator by 5 eliminates the $\frac{1}{5}$.
Before fractions can be **added** or **subtracted** they must have the same denominator:

$$4\tfrac{3}{5} + 2\tfrac{1}{4} = 6 + \frac{3}{5} + \frac{1}{4} = 6 + \frac{3 \times 4}{5 \times 4} + \frac{1 \times 5}{4 \times 5} = 6 + \frac{12}{20} + \frac{5}{20} = 6\tfrac{17}{20}$$

Add the whole numbers. The common denominator is the smallest number that 5 and 4 go into, i.e. 20. Express the fractions in terms of $\frac{1}{20}$. The fractions can now be added.

To **multiply** fractions, multiply the numerators together and multiply the denominators together.

$$4\tfrac{1}{2} \times 2\tfrac{2}{3} = \frac{9}{2} \times \frac{8}{3} = \frac{9 \times 8}{2 \times 3} = \frac{3 \times 4}{1 \times 1} = 12$$

Change to improper fractions where necessary. Multiply the numerators and denominators. Cancel where possible.

$\frac{1}{2}$ goes into 2 four times, i.e $2 \div \frac{1}{2} = 4$. To **divide** by a fraction invert the fraction and change the $\div$ to $\times$.

$$3\tfrac{1}{5} \div 1\tfrac{1}{7} = \frac{16}{5} \div \frac{8}{7} = \frac{16}{5} \times \frac{7}{8} = \frac{16 \times 7}{5 \times 8} = \frac{2 \times 7}{5 \times 1} = \frac{14}{5} = 2\tfrac{4}{5}.$$

Change to improper fractions, invert the second fraction and change the $\div$ to $\times$. Now proceed as in ordinary multiplication.

To evaluate a compound fraction:

$$\frac{\frac{3}{4}}{\frac{2}{3}} = \frac{3}{4} \div \frac{2}{3} = \frac{3}{4} \times \frac{3}{2} = \frac{9}{8} = 1\tfrac{1}{8}.$$

Decimals

A **decimal** is a fraction with its denominator in powers of 10.

$0{\cdot}24 = \frac{2}{10} + \frac{4}{100} = \frac{24}{100}$; $0{\cdot}02 = \frac{2}{100}$; $0{\cdot}035 = \frac{35}{1000}$.

The number of 0's in the denominator is equal to the number of digits in the decimal, after the decimal point.

To change a fraction into a decimal, divide the denominator into the numerator.

$$8 \,\underline{|\; 3{\cdot}000}$$
$$0{\cdot}375 \qquad \therefore \tfrac{3}{8} = 0{\cdot}375$$

It is well worth knowing the following:

$\frac{1}{2} = 0.5$: $\frac{1}{4} = 0.25$: $\frac{1}{5} = 0.2$: $\frac{1}{8} = 0.125$: $\frac{1}{20} = 0.05$

To multiply a decimal by 10, 100, 1 000, etc., move its decimal point 1, 2, 3, etc. places to the right, e.g. $4{\cdot}63 \times 100 = 463$ and $42{\cdot}5 \times 1\,000 = 42\,500$. To divide a decimal by 10, 100, 1 000 etc. move its decimal point 1, 2, 3, etc. places to the left, e.g. $46{\cdot}3 \div 10 = 4{\cdot}63$: $87 \div 100 = 0{\cdot}87$.

Rounding off

To **round off** a decimal to a given number of **decimal places** we proceed as follows: if the last figure is more than 5 increase the rounded-off digit by 1. If the last digit is less than 5 leave the rounded-off digit alone. If the last digit is 5, there is no real rule but in statistics it is becoming the habit to make the rounded-off digit even.

3·065 827 is a decimal expressed to 6 decimal places.
= 3·065 83 to 5 decimal places. The 7 changes the 2 to 3.
= 3·065 8 to 4 decimal places. The 3 leaves 8 unchanged.
= 3·066 to 3 decimal places.
= 3·07 to 2 decimal places and 3·1 to 1 decimal place.

To round off a number to a given number of **significant figures**, count the total number of digits in the result.
4·760 9 is a decimal expressed to 5 significant figures.
= 4·761 to 4 significant figures. The 9 changes 0 to 1.
= 4·76 to 3 significant figures. The 1 leaves the 6 unchanged.
= 4·8 to 2 significant figures.
= 5 to 1 significant figure or in this case, to the nearest whole number.

0·008 406 is a decimal expressed to 4 significant figures. It is important to realise that the 0's between the decimal point and the first non-zero digit are **not** significant.
∴ 0·008 41 is expressed to 3 significant figures and 0·008 4 to 2 significant figures.

In the case of whole numbers, consider 43 582, which might be the population of a town.
As 43 580 it is expressed to 4 significant figures.
As 43 600 it is expressed to 3 significant figures.
As 44 000 it is expressed to 2 significant figures. Do not omit the zeros. We are likely to say that the population is approximately 44 000.

When dividing and giving the answer to a certain number of decimal places, it is necessary to go to one more place than is required in order to round off. E.g. Evaluate 4·649 ÷ 11 to 3 decimal places.

$$11 \,\underline{|\; 4{\cdot}649}$$
$$0{\cdot}4226 \qquad 0{\cdot}423 \text{ to 3 decimal places.}$$

Dividing 3 into 1 we obtain the **recurring** decimal $\frac{1}{3} = 0{\cdot}333\,333\ldots$
This is written $0{\cdot}\dot{3}$. Similarly $\frac{2}{3} = 0{\cdot}\dot{6}$ and $\frac{1}{9} = 0{\cdot}\dot{1}$.
$\frac{1}{7} = 0{\cdot}142\,857\,142\,857\ldots$ the block of digits 142 857 recurring.
This is written $0{\cdot}\dot{1}42\,85\dot{7}$.

A recurring decimal can be rounded off in the same way as any decimal. E.g. $\frac{1}{3} = 0{\cdot}333$ to 3 decimal places or to 3 significant figures. But $\frac{2}{3} = 0{\cdot}667$ to 3 decimal places. The last 6 changes the rounded-off digit to 7. It is incorrect to call $\frac{2}{3} = 0{\cdot}666$.

Limits of accuracy

When the length of a line is 5 cm to the nearest cm, it is assumed that the length can lie anywhere between the values 4·5 and 5·5 cm, because any number between these values rounds off to 5. Such an expression is sometimes written $5 \pm 0{\cdot}5$ cm. Similarly, if the mass of an object is found to be 8·5 grams to the nearest 0·1 gram, the mass can be any number between 8·45 and 8·55 grams, i.e. $8{\cdot}5 \pm 0{\cdot}05$ grams. These are the **limits of accuracy** of the number 8·5. If an error is made, the maximum it can be is 0·05, whether too large or too small. This error can be expressed as a percentage as follows:

$$\% \text{ error} = \frac{\text{error}}{\text{mass}} \times 100 = \frac{0{\cdot}05}{8{\cdot}5} \times 100 = \frac{5}{8{\cdot}5}\%.$$

The % error = 0·588%

Example The length and breadth of a rectangle are measured as 8 cm and 6 cm to the nearest cm. What are the limits in which the calculated area must lie?

The length 8 cm can lie between 8·5 and 7·5 cm.
The breadth 6 cm can lie between 6·5 and 5·5 cm.
The smallest area calculated occurs when the dimensions are least,
$\therefore$ the smallest area $= 7{\cdot}5 \times 5{\cdot}5 = 41{\cdot}25 \text{ cm}^2$.
The largest area possible requires the largest dimensions.
$\therefore$ the largest area $= 8{\cdot}5 \times 6{\cdot}5 = 55{\cdot}25 \text{ cm}^2$.
The area lies between the limits 41·25 and 55·25 cm^2.

Example $x = 5{\cdot}4$ and $y = 3{\cdot}6$, both numbers given to 2 significant figures. Calculate (i) the maximum value of $x + y$ and (ii) the maximum value of $x - y$.

(i) x can lie between 5·45 and 5·35; y can lie between 3·65 and 3·55. The maximum $x + y$ = largest x + largest y = $5{\cdot}45 + 3{\cdot}65 = 9{\cdot}10$.

(ii) The maximum $x - y$ = largest x − smallest y (in order to obtain the largest difference) $= 5{\cdot}45 - 3{\cdot}55 = 1{\cdot}90$.

Indices

On page 17 the laws of **indices** are stated in algebraic terms. In arithmetical terms they can be summarised by stating that:

$3^2 \times 3^3 = 3^5 = 243$; $2^5 \div 2^3 = 2^2 = 4$; $(2^3)^2 = 2^6 = 64$.

Now consider $a^4 \div a^4 = a^{4-4} = a^0$. What does this mean?

Any number divided by itself = 1 $\therefore a^4 \div a^4 = 1 \;\therefore a^0 = 1$.
Also $a^2 \div a^3 = a^{2-3} = a^{-1}$. What does this mean?

$$a^2 \div a^3 = a^{2-3} = \frac{a \times a}{a \times a \times a} = \frac{1}{a} \;\therefore a^{-1} = \frac{1}{a} \text{ also } a^{-2} = \frac{1}{a^2}.$$

E.g. $7^0 = 8^0 = 1$; $10^{-1} = \frac{1}{10}$; $10^{-2} = 0{\cdot}01$; $3^{-1} = \frac{1}{3}$; $2^{-3} = \frac{1}{8}$.
The square root of x^6 is x^3. Also $\sqrt{x^4} = x^2$ and $\sqrt{x^2} = x$. In each case the index is divided by 2.
$\therefore \sqrt{x} = x^{\frac{1}{2}}$ similarly $\sqrt[3]{x} = x^{\frac{1}{3}}$ and $\sqrt[4]{x} = x^{\frac{1}{4}}$.
E.g. $4^{\frac{1}{2}} = 2$; $8^{\frac{1}{3}} = 2$; $81^{\frac{1}{4}} = 3$; $144^{\frac{1}{2}} = 12$; $10^{\frac{1}{2}} = 3{\cdot}162$.

Applying the laws of indices we obtain results such as:

$10^{-6} \times 10^8 = 10^2$ (adding indices); $a^4 \div a^{-2} = a^{4-(-2)} = a^6$.

The index notation can be extended to include more than one process, e.g. $8^{\frac{2}{3}} = (8^{\frac{1}{3}})^2 = 2^2 = 4$ also $81^{\frac{3}{4}} = (81^{\frac{1}{4}})^3 = 3^3 = 27$.
The index $\frac{2}{3}$ means 'take the cube root then square', and further to that $8^{-\frac{2}{3}} = \frac{1}{8^{\frac{2}{3}}} = \frac{1}{4}$. The minus sign means 'over'.

Standard form

A number expressed in the form $A \times 10^n$, where A is a number between 1 and 10 and n is a positive or negative integer, is written in **standard form** or **scientific notation**. This form is particularly useful for expressing very large or small numbers.

E.g. $3\,200\,000\,000 = 3{\cdot}2 \times 10^9$ in standard form. To obtain this:
(1) Place the decimal point between the first and second digits so that $A = 3{\cdot}2$.
(2) Count the number of decimal place moves to the **right** required, to restore the decimal point to its original position. It is 9.
$\therefore$ 3·2 is multiplied by 10^9. Hence the number $= 3{\cdot}2 \times 10^9$.

The very small number $0{\cdot}000\,000\,043 = 4{\cdot}3 \times 10^{-8}$. To obtain this:
(1) Proceed as before to make $A = 4{\cdot}3$.
(2) To restore the decimal point to its original position it must be moved 8 places to the **left**. 4·3 is divided by 10^8 or multiplied by 10^{-8}. Hence the number $= 4{\cdot}3 \times 10^{-8}$.

E.g. $4\,360\,000 = 4{\cdot}36 \times 10^6$ (4·36 million) and $0{\cdot}000\,743 = 7{\cdot}43 \times 10^{-4}$.
Numbers less than 1 always have a **negative** power of 10.

To evaluate $\dfrac{2{\cdot}4 \times 10^6 \times 2{\cdot}0 \times 10^3}{9{\cdot}6 \times 10^{-3}}$: $\left(\dfrac{2{\cdot}4 \times 2{\cdot}0}{9{\cdot}6} = \dfrac{4{\cdot}8}{9{\cdot}6} = 0{\cdot}5\right)$

$$= \frac{0{\cdot}5 \times 10^6 \times 10^3}{10^{-3}} = \frac{0{\cdot}5 \times 10^9}{10^{-3}} = 0{\cdot}5 \times 10^{9-(-3)} = 0{\cdot}5 \times 10^{12}$$

$= 5{\cdot}0 \times 10^{-1} \times 10^{12} = 5{\cdot}0 \times 10^{11}$ in standard form.

Evaluate the non-10 numbers first. Notice the use of the index laws.

Methods of calculation

In all examinations at this level, 3-or 4-figure tables are issued. Calculations can be done with the help of these, but it is generally accepted that slide-rule accuracy is sufficient.

It is essential to know what degree of accuracy is required in a particular question. Should the answer be exact, or to one or two significant figures? It is fair to say that in everyday life people like to be charged the correct fare, premium or cost for an article, therefore money questions should be calculated without logarithms or slide rules. In many questions the examiners will specify the accuracy required. Adhere to this, otherwise marks may be deducted.

Make a rough estimate before doing the actual calculation. Round some figures up and others down, e.g. $\frac{56 \times 8{\cdot}3}{3{\cdot}6} \approx \frac{60 \times 8}{4} \approx 120.$

The trigonometrical tables are explained on page 157. The logarithms, squares, roots and reciprocals under scrutiny here are all read in a similar way, except that in the 4-figure editions of tables the differences are **subtracted** in the case of the reciprocals.

Tables of squares

The 'squares' table does not indicate the decimal-point position. Use the following type of square to help make an estimate.

$20^2 = 400 : 40^2 = 1\,600 : 70^2 = 4900 : 300^2 = 90000 : 500^2 = 250000$

i.e. square the non-zero digit and **double** the number of **zeros**. To find $(57{\cdot}64)^2$ using the tables, locate the required square (at the intersection of line 57, column 6); at the intersection of line 57 and difference column 4 is found the adjustment to **add** on for the fourth figure. See the table below:

	6	4
57	3318	5

Hence $(57{\cdot}64)^2 = 3\,318 + 0005 = 3\,323$. These are the digits required for the answer. To find the decimal-point position make an estimate: $50^2 = 2\,500$ and $60^2 = 3\,600$ $\therefore$ $(57{\cdot}64)^2 = 3\,323{\cdot}0$ between the two values. Verify that $(23{\cdot}52)^2 = 553{\cdot}2$ and $(153)^2 = 23\,410$.

In the case of the square of a number less than 1 the following squares should be appreciated:

$(0{\cdot}1)^2 = 0{\cdot}1 \times 0{\cdot}1 = 0{\cdot}01$: $(0{\cdot}2)^2 = 0{\cdot}04$: $(0{\cdot}4)^2 = 0{\cdot}16\ldots$
$(0{\cdot}01)^2 = 0{\cdot}000\,1$: $(0{\cdot}02)^2 = 0{\cdot}000\,4$: $(0{\cdot}04)^2 = 0{\cdot}001\,6\ldots$

Square the non-zero digit and double the number of decimal places. Note that $(0{\cdot}3)^2 \neq 0{\cdot}9$. It is 0·09. Similarly $(0{\cdot}2)^2 = 0{\cdot}04$.

To find the value of $(0{\cdot}634\,2)^2$: line 63, column 4, and a difference of 2 gives the digits $4\,020 + 0\,003 = 4\,023$. Make the estimate $(0{\cdot}6)^2 = 0{\cdot}36$ and $(0{\cdot}7)^2 = 0{\cdot}49$. ∴ we can assume that $(0{\cdot}634\,2)^2 = 0{\cdot}402\,3$, between these two values.
Verify that $(0{\cdot}027)^2 = 0{\cdot}000\,729$ and $(0{\cdot}234)^2 = 0{\cdot}054\,8$ to 3 significant figures.

Notice that the square of a number less than 1 is smaller than the number itself. It is also worth noting that the 4-figure tables give the **exact** square of a **two-digit number**, i.e. the squares in the first column of the tables are exact.

Square-root tables

$\sqrt{36} = 6$ and $\sqrt{49} = 7$ ∴ $\sqrt{40}$ lies between 6 and 7. From the tables it will be found that $\sqrt{40}$ is 6·325. The fact that $\sqrt{4} = 2$ and $\sqrt{40} = 6{\cdot}325$ means that in the square-root tables, on line 40, column 0, we can find the two numbers 2 000 and 6 325. In some tables the roots of numbers between 1 and 10 are printed separately from the roots of numbers between 10 and 100, while in other tables the two possible roots are printed together on line 40 column 0. The difference is not difficult to see here, but how are we to find the square root of a number outside the range 1 to 100?

To find $\sqrt{5\,840}$:
the tables give the two numbers 2 417 and 7 642 on line 58, column 4. To select the correct number, proceed as follows:
(1) Write down the number including its decimal point. Pair off the digits from the decimal point. |58|40|·0
(2) Find the square root of the nearest square number **below** the first pair, 58. It is 7 ($7^2 = 49$).
(3) This indicates the first figure of the correct root. ∴ accept the number 7642 as the correct root required.
(4) Place the digits of the root above the pairs of the number:

7	6	·4 2
58	40	·0

Because there are two pairs of digits in the number there are two digits before the decimal point in the root $\therefore \sqrt{5\,840} = 76{\cdot}42$.

To find $\sqrt{584}$, pairing off as above gives |5|84|·0. The nearest root below 5 is $2(2^2 = 4)$. The required root is **2**417. Spacing them above the pairs gives $\sqrt{584} = 24{\cdot}17$.
Verify that $\sqrt{8\,531} = 92{\cdot}36$ and $\sqrt{853{\cdot}1} = 29{\cdot}21$.

For the square roots of numbers less than 1, proceed in a similar way. E.g. to find $\sqrt{0{\cdot}674\,1}$.
The tables offer the numbers 2 596 and 8 211 as the possible roots on line 67, column 4, with a difference of 1.
(1) Pair off from the decimal point 0·|67|41| in the opposite direction in this case.
(2) The nearest square root below 67 is 8 $(8^2 = 64)$
(3) The required root is **8** 211
(4) Space out the digits as before

	8	2	1 1
0·	67	41	

gives $\sqrt{0{\cdot}674\,1} = 0{\cdot}821\,1$.

To find $\sqrt{0{\cdot}009}$, the tables offer 3 000 and 9 487. Pair off as before 0·|00|90|. The nearest square root below 90 is 9 $(9^2 = 81)$. The required root is **9** 487. When the digits are spaced out over the pairs $\sqrt{0{\cdot}009} = 0{\cdot}094\,87$, the two 0's in the number producing one in the root.

The square root of a number less than 1 is greater than the number. This method of finding a root is the initial step in a method for finding roots without tables. Used here it gives the first figure of the root, enough to make the right selection.

Reciprocals

The **reciprocal** of 2 is $\frac{1}{2}$. It appears in the table as 5 000. The reciprocal table gives a four-figure number which can be used to represent '1 over' a number. Hence the reciprocal of 3 is 0·333 3. The decimal point is not given, so it is necessary to make an estimate. Remember to **subtract** the difference.

To find the reciprocal of 23·45: the reciprocal table gives $4\,274 - 9 = 4\,265$. To place the decimal point make an estimate $\frac{1}{2} = 0{\cdot}5$, $\frac{1}{20} = 0{\cdot}05$. $\dfrac{1}{23{\cdot}45}$ is somewhat less than this $\therefore$ the required value is 0·042 65 or **0·043** to 2 significant figures.

To evaluate $\frac{1}{0{\cdot}073}$: the table gives the number 1370.

Estimate: $\frac{1}{0{\cdot}07} \approx \frac{100}{7} \approx 14 \therefore \frac{1}{0{\cdot}073} \approx 13{\cdot}70.$

Logarithms

A **logarithm** or log is an **index** or a **power**. $2^3 = 8$ states that 3 is the log of 8 when expressed in powers of 2. Similarly 4 is the log of 81 when in powers of 3 because $3^4 = 81$. The **common logs** used for calculations express numbers as powers of 10. To find the log of 2 look along line 20, column 0, to find the number 3010. This means that $2 = 10^{0{\cdot}3010}$ or that $\log 2 = 0{\cdot}3010$.
Similarly $4{\cdot}36 = 10^{0{\cdot}6395}$ or $\log 4{\cdot}36 = 0{\cdot}6395$.
To find the log of 436 standard form can be used.

$$436 = 4{\cdot}36 \times 10^2 = 10^{0{\cdot}6395} \times 10^2 = 10^{2{\cdot}6395} \therefore \log 436 = 2{\cdot}6395$$

The 2 before the decimal point is called the **characteristic** of the log and registers the power of 10. Hence the log $43{\cdot}6 = 1{\cdot}6395$ and $\log 4\,360 = 3{\cdot}6395$.

To find the log of a number less than 1, e.g. 0·043 6. Writing it in standard form, $0{\cdot}043\,6 = 4{\cdot}36 \times 10^{-2}$. The power of 10 is -2 and we write $\log 0{\cdot}043\,6 = \bar{2}{\cdot}6395$. The characteristic is called bar 2. $\text{Log } 0{\cdot}000\,436 = \bar{4}{\cdot}6395$ and $\log 0{\cdot}004\,36 = \bar{3}{\cdot}6395$.

Notice that $\log 100 = 2{\cdot}0000$ and $\log 1\,000 = 3{\cdot}0000$.

To find the number represented by a log, e.g. 0·5732, look in the **anti-log** table. Look along the line 57, column 3, with a difference of 2 gives $3741 + 2 = 3743$. The characteristic 0 indicates that the number lies between 1 and 10 $\therefore$ 3·743 is the anti-log.

The anti-log of $2{\cdot}5732 = 374{\cdot}3$. The characteristic 2 indicates that 3·743 is multiplied by 10^2 or 100.

The anti-log of $\bar{1}{\cdot}5732$ is 0·3743. The characteristic $\bar{1}$ indicates that 3·743 should be multiplied by 10^{-1}, i.e. divided by 10.

Logarithms are indices, therefore to **multiply** numbers **add** the logs and to **divide** numbers **subtract** the logs.

Evaluate to 3 significant figures $\frac{468 \times 0{\cdot}748}{16{\cdot}24}$

Estimate: $\approx \dfrac{500 \times 1}{20} \approx 25$

	Number	Log	
	468	2·6702	
×	0·748	$\bar{1}$·8739	+
		2·5441	
÷	16·24	1·2105	−
	2156	1·3336	

Answer: 21·6 (consistent with estimate).

Draw the framework as shown, write down the numbers find the logs and work out each characteristic carefully. **Add** the logs of the numerator to give 2·5441 and from that **subtract** the log of the denominator. Find the anti-log of ·3336. The characteristic is 1 ∴ the answer is 21·56, or 21·6 to 3 significant figures.

To find squares and square roots of numbers, use square and root tables. Other powers could require logs. E.g. To find the value of $(0{\cdot}346)^3$ to 3 significant figures: from the laws of indices **multiply** the log by 3.

Estimate: $0{\cdot}3 \times 0{\cdot}3 \times 0{\cdot}3 \approx 0{\cdot}09 \times 0{\cdot}3 \approx 0{\cdot}027$

Number	Log	
0·3460	$\bar{1}$·5391	
	3	×
4143	$\bar{2}$·6173	

Answer: 0·041 4 (consistent with estimate).

Multiply the log by 3. $\bar{1} \times 3 = \bar{3}$ with 1 to carry makes the final characteristic $\bar{2}$. The answer is $4{\cdot}143 \times 10^{-2} = 0{\cdot}041\,43$.

To find $\sqrt[3]{21{\cdot}6}$. From the law of indices, the cube root is power $\frac{1}{3}$, ∴ **divide** the log by 3. Verify that the result is 2·785.

In the case of $\sqrt[3]{0{\cdot}641}$, the log of 0·641 = $\bar{1}$·8069. The bar number is not a multiple of 3. Write the log as $\bar{3}+2{\cdot}8069$ and divide thus:

$$3 \;\underline{|\; \bar{3}+2{\cdot}8069}$$
$$\bar{1}{\cdot}9356$$

Divide 3 into $\bar{3}$ then divide 3 into 28.

Anti-logging gives 8 622, the characteristic $\bar{1}$ means that $\sqrt[3]{0{\cdot}641} =$ 0·862 2.

The trickiest part of logarithms is dealing with bar numbers. Some examples are given here to show how to add and substract.

$$\text{(i)}\quad \begin{array}{r} 2{\cdot}4000 \\ +\bar{3}{\cdot}7000 \\ \hline 0{\cdot}1000 \\ \hline \end{array} \qquad \text{(ii)}\quad \begin{array}{r} 2{\cdot}4000 \\ -\bar{3}{\cdot}7000 \\ \hline 4{\cdot}7000 \\ \hline \end{array} \qquad \text{(iii)}\quad \begin{array}{r} \bar{1}{\cdot}6000 \\ -2{\cdot}7000 \\ \hline \bar{4}{\cdot}9000 \\ \hline \end{array} \qquad \text{(iv)}\quad \begin{array}{r} \bar{1}{\cdot}5000 \\ +\bar{2}{\cdot}6000 \\ \hline \bar{2}{\cdot}1000 \\ \hline \end{array}$$

In (i) with 1 to carry we have $2+(-3+1) = 2-2 = 0$
In (ii) with 1 to carry we have $2-(-3+1) = 2-(-2) = 4$
The reader is left to verify (iii) and (iv).

The slide rule

Each slide rule carries its own set of instructions, but it is possible to provide a general guide for using one to multiply and divide numbers. To carry out these processes use the *C* and *D* scales.

C 1 2 3 4 5 6 7 8 9 10
D

This scale is the **logarithm** scale. The distance between 1 and 2 represents log 2, the distance between 1 and 3 is log 3, etc. A section of the scale is produced here to show how to read off the numbers. $P = 4{\cdot}2$, $Q = 4{\cdot}65$ and $R = 4{\cdot}88$.

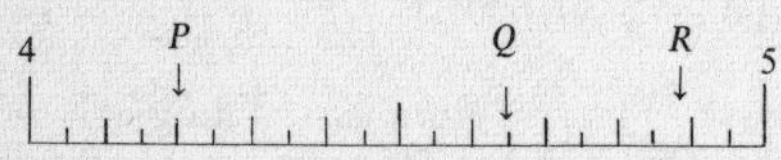

To evaluate $1{\cdot}5 \times 42$, see figure 61(a).
(1) Put the 1 of the *C* scale above 15 of the *D* scale.
(2) Find 42 on the *C* scale. The required answer appears on the *D* scale immediately below this. It reads 63.
(3) Make an estimate: $1 \times 42 = 42$;
the required answer is 63·0.

We have **added** the 14 distance on the *D* scale to the 26 distance on the *C* scale.

To evaluate $1{\cdot}4 \times 820$, see figure 61(b).
Proceeding as before, the 82 of the *C* scale projects beyond the *D* scale and gives no reading. Proceed as follows:
(1) Put the 10 of the *C* scale over the 82 of the *D* scale.
(2) Find 14 on the *C* scale. The required answer appears immediately below this on the *D* scale. It reads 115.
(3) Make an estimate: $2 \times 800 = 1\,600$;
the required answer is 1 150.

To evaluate $3{\cdot}4 \div 1{\cdot}8$, see figure 61(c).
(1) Put the 18 of the *C* scale over the 34 of the *D* scale.

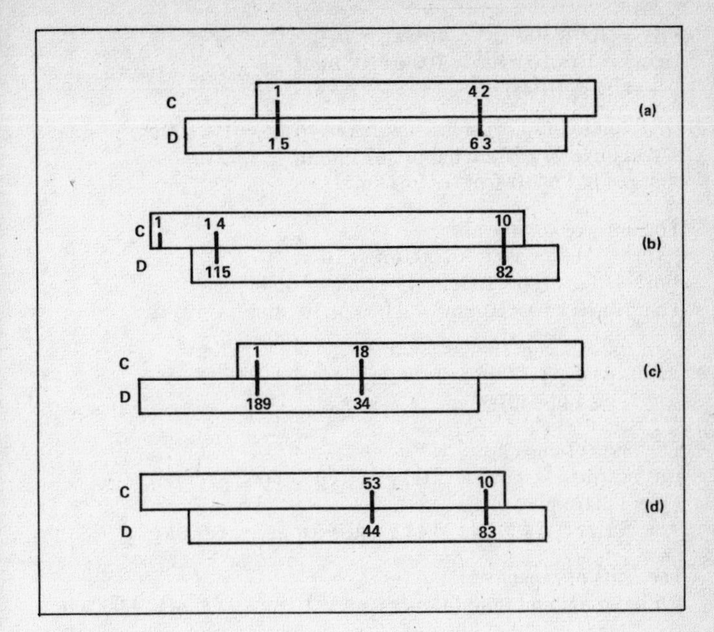

Figure 61

(2) Find the 1 on the *C* scale. The answer appears below it on the *D* scale. It reads 189.
(3) Make an estimate: $3 \div 2 = 1{\cdot}5$;
the required answer is 1·89.

To evaluate $4{\cdot}4 \div 53$, see figure 61(d).
(1) Put the 53 of the *C* scale over the 48 on the *D* scale.
(2) The 1 of the *C* scale projects beyond the *D* scale. Now find the 10 of the *C* scale. The required answer appears below it on the *D* scale. It reads 830.
(3) Make an estimate: $4 \div 50 = 8 \div 100 = 0{\cdot}08$;
the required answer is 0·083.

The metric system

The units of length are:

1 km = 1 000 m: 1 m = 100 cm: 1 cm = 10 mm
∴ 462 cm = 4·62 m: 5·4 km = 5 400 m: 86 mm = 8·6 cm.

The units of area are:
$1\,\text{km}^2 = 1\,000 \times 1\,000\,\text{m}^2 = 1$ million or $10^6\,\text{m}^2$.

$1\ m^2 = 100 \times 100\ cm^2 = 10\,000$ or $10^4\ cm^2$.
$1\ cm^2 = 10 \times 10\ mm^2 = 100$ or $10^2\ mm^2$.
1 hectare $= 10\,000\ m^2$.

$\therefore$ a square of side 4 m has an area of $16\ m^2 = 160\,000\ cm^2$.
A rectangle 90×80 cm has area $7\,200\ cm^2 = 0{\cdot}72\ m^2$.
An area of $1\,640\,000\ m^2 = 1{\cdot}64\ km^2$.

The units of volume are:
$1\ km^3 = 1\,000 \times 1\,000 \times 1\,000\ m^3 = 10^9\ m^3$.
$1\ m^3 = 100 \times 100 \times 100\ cm^3 = 1$ million or $10^6\ cm^3$.
$1\ cm^3 = 10 \times 10 \times 10\ mm^3 = 1\,000$ or $10^3\ mm^3$.

$\therefore$ a cube of side 2 m has volume $8\ m^3 = 8 \times 10^6\ cm^3$.
A cuboid $4 \times 3 \times 2$ cm has volume $24\ cm^3 = 24\,000\ mm^3$.
A volume of $6\,340\,000\ cm^3 = 6{\cdot}34\ m^3$.

The units of mass are:
1 tonne (metric ton) = 1 000 kg: 1 kg = 1 000 g (gram):
1 g = 1 000 mg,
$\therefore$ a mass of 35 400 g = 35·4 kg and 5·6 tonne = 5 600 kg.

The units of capacity are:
1 litre = 1 000 ml (millilitre): 1 ml = 1 cm^3: 1 litre = 1 000 cm^3.

It is worth remembering that the mass of 1 cm^3 of water is 1 g (at 4° C), $\therefore$ the mass of 1 litre of water is 1 kg.

Remember that 'kilo-' stands for 1 000 and 'milli-' stands for $\frac{1}{1000}$.

Mistakes are often made in converting units of area or volume. E.g. when converting cm^2 to m^2, or *vice versa*, the square of the conversion number is required. $\therefore$ for cm^2 to m^2, $\div 100^2$ and for m^2 to cm^2, $\times 100^2$. Similarly, to convert cm^3 to m^3, $\div 100^3$.

The following formulae are useful for calculation of areas and volumes:

Triangle: area $= \frac{1}{2}bh$
Trapezium: area $= \frac{1}{2}(a+b)h$
Parallelogram: area $= bh$
Circle: area $= \pi r^2$, circumference $= 2\pi r$
Sphere: volume $= \frac{4}{3}\pi r^3$, curved surface area $= 4\pi r^2$
Cylinder: volume $= \pi r^2 h$, curved surface area $= 2\pi rh$, total area $= 2\pi r^2 + 2\pi rh$
Cone: volume $= \frac{1}{3}\pi r^2 h$, curved surface area $= \pi r \times$ (slant height)

Cuboid: volume $= lbh$
Pyramid: volume $= \frac{1}{3}(\text{base area})h$
Prism: volume $= (\text{base area})h$

The value of π is usually taken as 3·142, 3·14 or $\frac{22}{7}$ if 7 cancels.

A **prism** is a solid with uniform cross-section. Cuboids and cylinders are prisms. **The volume of a prism = (area of cross-section) × (perpendicular height).** Hence the volume of a cylinder is $(\pi r^2)h$. Notice that the volume of a pyramid is a third the volume of a cuboid with the same cross-section and height. Hence the volume of a cone $= \frac{1}{3}\pi r^2 h$.

Two problems including a selection of these formulae are included here. The calculations need only be to slide-rule accuracy. Make an estimate to ensure that the answer is sensible.

Example Figure 62(a) shows the cross-section of a swimming pool, with water 2·5 m deep one end and 1 m deep the other. The pool is 33 m long and 12 m wide. Find (i) the volume of water needed to fill it; (ii) the mass of the water in kg.

(i) The solid formed is a prism with cross-section a trapezium.
Area of cross-section $= \frac{1}{2}(2{\cdot}5+1) \times 33\,\text{m}^2$
$\therefore$ the volume of water $= \frac{1}{2}(2{\cdot}5+1) \times 33 \times 12\,\text{m}^3$(cross-section × height) $= 3{\cdot}5 \times 33 \times 6\,\text{m}^3 = 21 \times 33 = 693\,\text{m}^3$ of water.
Change this to $\text{cm}^3 = 693 \times 100 \times 100 \times 100$
Change this to litres $= 693 \times 100 \times 10\ (\div 1000)$
Number of litres required $= 693\,000$ or $6{\cdot}93 \times 10^5$.
(ii) the mass $= 693\,000$ kg. (The mass of 1 litre of water is 1 kg.)

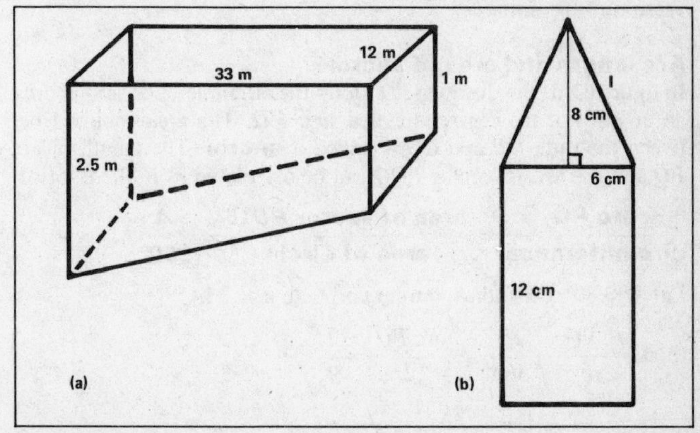

Figure 62

Example Figure 62(b) shows the cross-section of a solid comprising a cone surmounting a cylinder. The heights of the cone and cylinder are 8 and 12 cm respectively. The common radius is 6 cm. Find (i) the total volume; (ii) the total surface area; (iii) the mass of the solid if the density of the metal used is 5·4 g/cm^3. ($\pi = 3{\cdot}14$) (iv) the solid is a model of a rocket made to a scale of 1 : 100. Find the actual volume.

Never evaluate π until absolutely necessary.

(i) Volume cylinder $= \pi r^2 h = \pi 6 \times 6 \times 12 = 432\pi$ cm^3;
volume cone $= \frac{1}{3}\pi r^2 h = \frac{1}{3}\pi 6 \times 6 \times 8 = \pi \times 2 \times 6 \times 8$
$= 96\pi$ cm^3;
total volume $= 432\pi + 96\pi = 528\pi = 1\,660$ cm^3 to 3 significant figures.

(ii) Curved area of cone $= \pi r \times$ slant height.
In the figure $l = 10$ cm. (This is a 3 : 4 : 5 triangle with sides $\times 2$.)
$\therefore$ curved area of cone $= \pi \times 6 \times 10 = 60\pi$ cm^2
The area of the cylinder showing $= \pi r^2 + 2\pi rh$
$= \pi 6^2 + 2\pi \times 6 \times 12 = 36\pi + 144\pi = 180\pi$ cm^2
$\therefore$ the total surface area $= 60\pi + 180\pi = 240\pi = 753{\cdot}6$ cm^2.

(iii) The mass $= 1\,660 \times 5{\cdot}4$ g $= 8\,964$ g or 8·96 kg to 3 significant figures.

(iv) Full-size volume $= 100 \times 100 \times 100 = 10^6$ times larger than the model, $\therefore$ the actual volume $= 1\,660 \times 10^6$ cm^3 or 1 660 m^3.

On considering this question carefully it cannot be said that the calculations are laborious. Most mistakes will be made in the conversion of the units.

Arc length and area of sector

In figure 63(a) the distance PQ along the circumference, subtending an angle A at the centre is called **arc *PQ***. The area enclosed between the radii OP and OQ is called a **sector**. The length of arc PQ and the area of sector POQ can be calculated from these ratios:

$$\frac{\textbf{arc } \boldsymbol{PQ}}{\textbf{circumference}} = \frac{\textbf{area of sector } \boldsymbol{POQ}}{\textbf{area of circle}} = \frac{\boldsymbol{A}}{\textbf{360°}}$$

Let $A = 80°$, the radius = 6 cm and take $\pi = 3{\cdot}14$,

then $\dfrac{\text{arc } PQ}{2\pi 6} = \dfrac{80°}{360°} \therefore \dfrac{\text{arc } PQ}{12\pi} = \dfrac{2}{9}$

arc $PQ = 12\pi \times \dfrac{2}{9} = \dfrac{8\pi}{3} = 8{\cdot}37$ cm.

$$\frac{\text{The area of sector } POQ}{\pi \times 6^2} = \frac{80}{360} \therefore \frac{\text{sector } POQ}{36\pi} = \frac{2}{9}$$

$$\therefore \text{ sector } POQ = \frac{2}{9} \times 36\pi = 8\pi$$

The area of the sector $POQ = 25{\cdot}12\,\text{cm}^2$

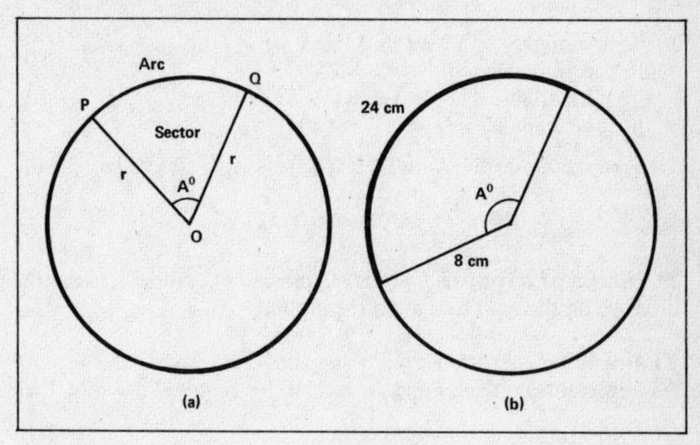

Figure 63

Example Figure 63(b) shows the plan of a cylindrical can of radius 8 cm. A label 24 cm wide is stuck to the can. What angle does it subtend at the centre?

Using the above ratio: $\dfrac{24}{2 \times \pi \times 8} = \dfrac{A}{360°}$

$$A = \frac{360 \times 24}{16\pi} = \frac{540}{\pi} = 172° \text{ to the nearest degree.}$$

Ratio

A **ratio** compares two or more quantities. E.g. The two prizes of £15 and £10 in a competition are in the ratio 15 : 10, which can be cancelled down to 3 : 2 as in fractions. In fact, a ratio is a fraction. $\dfrac{\text{1st prize}}{\text{2nd prize}} = \dfrac{15}{10} = \dfrac{3}{2}$. The first is $1\frac{1}{2}$ times larger than the second.

To divide £36 between A, B and C in the ratio 2 : 3 : 4.
Divide 36 into $2+3+4 = 9$ parts.
Each part will be $£36 \div 9 = £4$.

A receives £4 × 2 = £8; B receives £4 × 3 = £12; C receives £16. Check the answer by adding the shares to give £36.

Consider a pie chart in which the share-out of a quantity is in the ratio 3 : 5 : 6 : 10. Find the angles into which each sector is divided. The chart is to be divided into 3+5+6+10 = 24 parts.
Each part = 360° ÷ 24 = 15°

$\therefore$ the first angle = 15° × 3 = 45°
the second angle = 15° × 5 = 75°
the third angle = 15° × 6 = 90°
the fourth angle = 15° × 10 = 150°

Two masses are in the ratio 6 : 5. The smaller is 35 kg. To find the larger x: $\frac{x}{35} = \frac{6}{5} \Leftrightarrow x = \frac{6}{5} \times 35 \Leftrightarrow x = 42$ kg.

The **scale of a map** (e.g. 1 : 25 000) indicates the ratio between the distance on the map to that on the ground.

Example Let the distance on the map be 3·5 cm. To find the distance this represents on the ground, let x cm be the required distance, then

the ratio $\frac{\text{ground}}{\text{map}} = \frac{x}{3{\cdot}5} = \frac{25\,000}{1}$

$\Leftrightarrow x = 3{\cdot}5 \times 25\,000\text{ cm} = 87\,500\text{ cm} = 875\text{ m or } 0{\cdot}875\text{ km}.$

Let the distance on the ground be 4·5 km. To find the distance on the map, let y km be the distance on the map, then

$$\frac{\text{map}}{\text{ground}} = \frac{y}{4{\cdot}5} = \frac{1}{25\,000} \Leftrightarrow y = \frac{4{\cdot}5}{25\,000}\text{ km}.$$

$$y = \frac{4{\cdot}5 \times 1\,000 \times 100}{25\,000}\text{ cm} = 18\text{ cm}.$$

Example The area on the map is 6 cm². Find the area on the ground this represents.

The area on the ground = $6 \times (25\,000)^2\text{ cm}^2$. Change to standard form and **convert to m²**. This gives

$$\frac{6 \times 2{\cdot}5 \times 2{\cdot}5 \times 10^4 \times 10^4}{100 \times 100} = \frac{6 \times 6{\cdot}25 \times 10^8}{10^4} = 37{\cdot}5 \times 10^4\text{ m}^2$$

$$= 3{\cdot}75 \times 10^5\text{ m}^2.$$

Percentages

A **percentage** (%) is a fraction with denominator 100. $20\% = \frac{20}{100}$ and $250\% = \frac{250}{100}$ or 2·5. The following should be recognised immediately: $50\% = \frac{1}{2}$; $25\% = \frac{1}{4}$; $75\% = \frac{3}{4}$; $33\frac{1}{3}\% = \frac{1}{3}$; $10\% = \frac{1}{10}$.

To change a fraction to a percentage, multiply the fraction by 100 E.g. $\frac{3}{8}$ becomes $\frac{3}{8} \times 100 = \frac{3}{2} \times 25 = \frac{75}{2} = 37{\cdot}5\%$.

To change a decimal to a percentage, write the decimal with denominator 100. $0{\cdot}2 = \frac{20}{100}$ or 20%; $0{\cdot}35 = 35\%$; $0{\cdot}04 = 4\%$; $0{\cdot}015 = 1{\cdot}5\%$.

To find 15% of 43: we require $\frac{15}{100}$ of 43, i.e. $\frac{15}{100} \times 43 = \frac{645}{100} = 6{\cdot}45$. Do not automatically cancel. It is sometimes easier to preserve the 100 in the denominator. Multiply the numerators and then move the decimal point two places to the left.

To increase 55 by 12%: we require

$$55 + 55 \times \frac{12}{100} = 55\left(1 + \frac{12}{100}\right) \quad \text{(55 is a factor)}$$

This is $55\left(\frac{100}{100} + \frac{12}{100}\right) = 55 \times \frac{112}{100} = \frac{6160}{100} = 61{\cdot}60$.

(i) $\frac{100+r}{100}$ and (ii) $\frac{100-r}{100}$. (i) increases and (ii) decreases a number by $r\%$.

Example The length of a rectangle is increased by 30% and the breadth is decreased by 10%. Find the percentage change in the area.

Let the original area $A = lb$.

The new length $= \frac{130}{100}l$ the new breadth $= \frac{90}{100}b$

$$\text{the new area} = \frac{130}{100}l \times \frac{90}{100}b = \frac{13 \times 9}{100}lb = \frac{117}{100}lb$$

The new area is an increase of 17%.

Example A number when decreased by 22% becomes 39. Find the original number. Let x be the original number. When decreased by 22% it becomes:

$$\frac{78x}{100} \quad \therefore \quad \frac{78x}{100} = 39 \Leftrightarrow x = 39 \times \frac{100}{78} \Leftrightarrow x = \frac{100}{2},$$

so the number was 50.
The original number is 100%. The new number is 78%, ∴ 39 is increased in the ratio 100 : 78 to restore its value.

Percentages and money

1% of £1 = 1p so 15% of £4 = 60p, $46\frac{1}{2}$% of £2 = 93p. These are easy to see, but when the numbers are more difficult treat them as in an ordinary percentage sum.
E.g. To find $11\frac{1}{2}$% of £3·60:

$$\frac{11\frac{1}{2}}{100} \times 3{\cdot}60 = \frac{23}{200} \times 3{\cdot}60 = \frac{23 \times 1{\cdot}80}{100} = \frac{£41{\cdot}40}{100} = 41\tfrac{1}{2}\text{p}$$

E.g. An article is sold for £2·60 at a gain of 30% on the cost price. Find the cost price.
Let the cost price be £x. The selling price is an increase of 30% and is:

$$\frac{130x}{100} \quad \therefore \quad \frac{130x}{100} = 2{\cdot}60 \Leftrightarrow x = \frac{2{\cdot}60 \times 100}{130} \Leftrightarrow x = 200\text{p}.$$

The cost price is £2·00.

E.g. An article bought for £3·60 sells for £4·14. Find the profit percentage. The actual profit = £4·14 − £3·60 = 54p. Compare this with the cost price (360p) $\frac{54}{360}$ as a % $\frac{9}{60} \times 100 = 15\%$ profit.

Interest

A sum of money loaned or borrowed is the **principal** P.
The money added for this privilege is the **interest** I.
The percentage at which the interest is calculated is the **rate** R.
The time over which the interest is calculated is the time T.
The total repaid, principal + interest = the amount repaid.

Consider P = £50; R = 5% per annum; T = 3 years.
If the money is repaid with **simple interest**, the interest is calculated on the £50 each year.

$$I \text{ for 1 year} = 50 \times \frac{5}{100} \quad \therefore \text{ for 3 years} \quad \frac{50 \times 5 \times 3}{100} = \frac{750}{100} = £7{\cdot}50$$

The total amount repaid = £50 + £7·50 = £57·50

The simple interest formula $I = \dfrac{PRT}{100}$

If the money repaid with **compound interest**, the interest is calculated on the amount at the end of the previous year.

The amount after 1 year $= 50 \times \frac{105}{100} = \frac{525}{100} = £52{\cdot}50$ (increase of 5%).

The amount after 2 years $= 52{\cdot}50 \times \frac{105}{100} = £55{\cdot}125.$

The amount after 3 years $= 55{\cdot}125 \times \frac{105}{100} = \frac{£5788{\cdot}125}{100} = £57{\cdot}88.$

This can also be written $£50 \times 1{\cdot}05 \times 1{\cdot}05 \times 1{\cdot}05 = £50 \times 1{\cdot}05^3$.

Notice that the compound interest is dearer by 38p.

Key terms

The **factors** of a number are those numbers which divide into it exactly.
The **numerator** of a fraction is placed above the line, the **denominator** below the line.
The number 34·62 is expressed to 2 **decimal places** and to 4 **significant figures**.
All measurements are approximate. The **limits of accuracy** are the range in which a given measurement can lie. E.g. 4 cm $=4 \pm 0{\cdot}5$ cm.
Standard form or scientific notation is a number written in the form $A \times 10^n$, A between 1 and 10, n an integer.
A **logarithm** is an index. Logarithms follow the laws of indices.
The **characteristic** of a log denotes the power of 10 in the number.
In the metric system '**kilo-**' = 1000 and '**milli-**' = $\frac{1}{1000}$.
A **prism** is a solid with uniform cross-section.
The **sector** of a circle is a 'piece of cake', the area enclosed between two radii and the arc they cut off.
A **ratio** compares numbers. It is treated as a fraction.
A **percentage** (%) is a fraction with denominator 100.
Simple interest is calculated annually on the sum borrowed.
Compound interest is calculated annually on the amount at the end of the previous year.

Chapter 13
Statistics

Many decisions that affect our lives today are taken on the results of collected and analysed data or statistics. This data must be presented in a clear, concise way so that it is easily understood. For GCE O-level it is advisable to know the terms used and to be able to carry out the basic processes easily.

The pie chart

In a particular year in the U.K. the value of the output of crops was £320 million. The value of each crop is shown in figure 64(a) and with it is shown the **pie chart**, which is the best way of presenting this kind of table in pictorial form. It shows the share-out of the £320 million. The 360° of the circle are divided into the ratio of the sums of money, see figure 64(b).

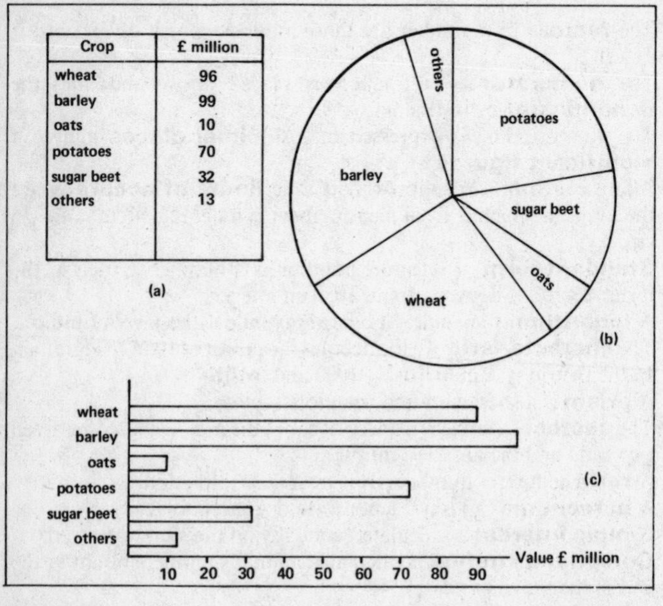

Crop	£ million
wheat	96
barley	99
oats	10
potatoes	70
sugar beet	32
others	13

Figure 64

The sector representing wheat has an angle $\frac{96}{320} \times 360 = 108°$ and that of oats is $\frac{10}{320} \times 360 = 11{\cdot}25°$. It is of little value as an aid to

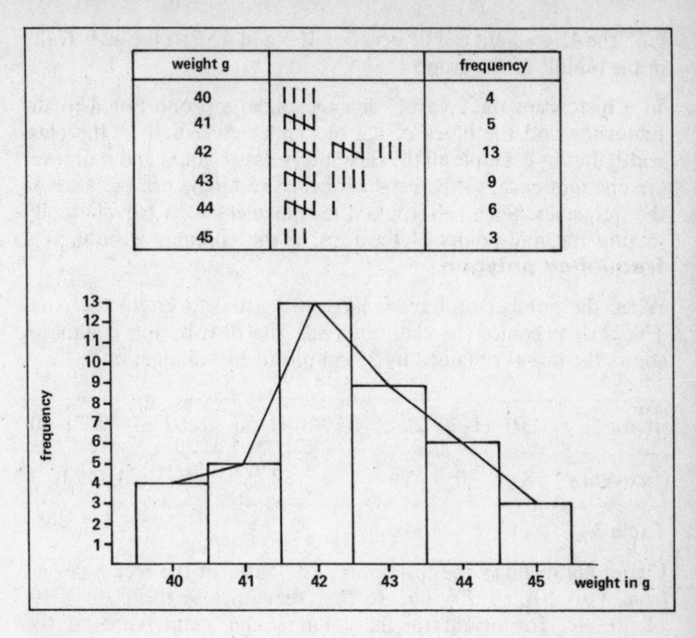

weight g		frequency
40	IIII	4
41	𝍸	5
42	𝍸 𝍸 III	13
43	𝍸 IIII	9
44	𝍸 I	6
45	III	3

Figure 65

calculation but one can see at a glance where the best values lie. The information can also be shown on a **bar chart**, see figure 64(c).

Frequency distribution

Each of a sample of 40 eggs is weighed to the nearest gram:

44 43 42 41 44 40 42 41 40 42 43 42 43 42 45 42
43 44 43 42 40 42 41 43 41 42 40 43 44 42 41 43
45 42 42 43 44 45 42 44.

To analyse these results, draw a table by taking a tally of them. Use a diagonal line to indicate every fifth item in each group.

The totals are written in the column headed 'frequency', which states the number of times that each weight occurs. Each weight is called a class. The class and frequency columns form a **frequency distribution**. Such a distribution must be shown on a **histogram** as in figure 65. The frequencies are shown on the vertical axis and the classes on the horizontal one. The class is written in the middle of the base of the column. In this case the weights are to the nearest

g. ∴ the 42 g weight can lie between 41·5 and 42·5 so the 42 is really in the middle of the group.

In a histogram the area of each rectangle is proportional to the frequency and the bases of the rectangles are equal to the class width. In this example all the class intervals are equal and moreover are one unit each, so the heights of each rectangle can be taken as the frequency. Such a histogram is equivalent to a bar chart. By joining the mid-points of the tops of the columns we obtain a **frequency polygon**.

When the number of items is large it is usual to group the data. The class becomes the class interval. The distribution in table 3 shows the marks obtained by 250 pupils in an examination:

mark	1–10	11–20	21–30	31–40	41–50	51–60	61–70	71–80
frequency	8	20	26	46	57	49	34	10

Table 3

Figure 66(a) shows the histogram. The bases of the rectangles go from $\frac{1}{2}$ to $10\frac{1}{2}$, to $20\frac{1}{2}$, etc. so that they enclose the class 1–10, 11–20, etc. The heights of the columns can again represent the frequencies because all the class intervals are equal.

mark	0–4	5–7	8–9
frequency	10	15	5

Table 4

When the class intervals are uneven as in the distribution in table 4, which shows the marks gained in a test by 30 pupils, the histogram appears as in figure 66(b). The area of each rectangle is proportional to its frequency so the 0–4 column which is 5 units wide will have height = 2 units so that the area is 10. Similarly the 5–7 column has width 3 units ∴ height = $\frac{15}{3}$ = 5 units and the 8–9 column has height $\frac{5}{2}$ = 2·5.

Averages
It is often necessary to compare two or more frequency distributions. In order to do this, it is useful to compare three particular characteristics of each distribution, the averages **mode**, **mean** and **median**.

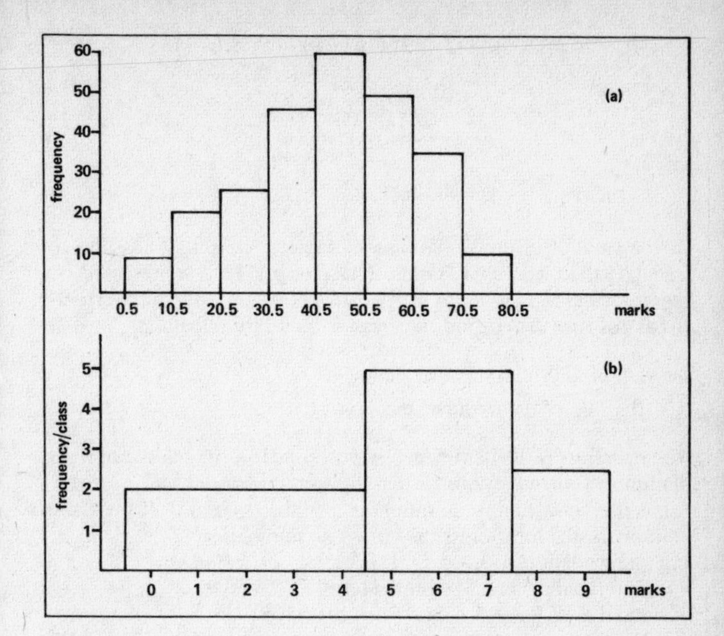

Figure 66

The **mode** is the class which has the highest frequency. E.g. it is a useful pointer to a manufacturer of clothes; he can see which of his products is selling the best. In figure 65 the mode is 42 g, in the grouped distribution in table 3 the **modal class** or **group** is 41–50 marks.

The **mean**, or what is commonly called the average, is calculated by $\frac{\text{the sum of the items}}{\text{number of items}}$

A pupil who scores 60, 32, 26, 51, 47 marks in 5 examinations has a mean score of $\frac{60+32+26+51+47}{5} = \frac{216}{5} = 43{\cdot}2$

To find the mean of a frequency distribution (as in figure 65 relating to the weights of eggs):

$$\text{the mean} = \frac{\text{total weight of eggs}}{\text{number of eggs}} = \frac{\text{sum of weight} \times \text{frequency}}{\text{sum of the frequencies}}$$

$$= \frac{(40\times 4)+(41\times 5)+(42\times 13)+(43\times 9)+(44\times 6)+(45\times 3)}{4+5+13+9+6+3}$$

$$= \frac{160+205+546+387+264+135}{40} = \frac{1697}{40}$$

$\therefore$ the mean = 42·4 grams.

In the case of grouped data as in table 3, it is not possible to calculate the mean as it stands. The group 1–10 is represented by the number $5\frac{1}{2}$, 11–20 by $15\frac{1}{2}$, etc. These are called the **mid-interval numbers**, and the mean is found by calculating

$$\frac{\text{the sum of mid-interval} \times \text{frequency}}{\text{the sum of the frequencies}}$$

The **median** is the item which is positioned in the middle of a set of numbers when they are written in ascending or descending order. It has the same number of numbers on either side of it. If there are n numbers, the location is the $\frac{1}{2}(n+1)$th number.
E.g. the median of 1, 3, 4, 5, 6, 11, 15 is 5
$n = 7$ so the median is the fourth term.
The median of 1, 3, 4, 5, 6, 11, 15, 19, 20, 30 is $8\frac{1}{2}$.
$n = 10$ an even number, so the median is the $\frac{1}{2}(10+1)$th term = $5\frac{1}{2}$th term. This is found between 6 and 11. Take the mean of them, $8\frac{1}{2}$, as the median.

To find the median of a frequency distribution as in figure 65. There are 40 weights, $\therefore$ the median is found at the $\frac{1}{2}(40+1) = 20\frac{1}{2}$th. Rather than write all the 40 weights in ascending order, proceed as follows: firstly, group the numbers in ascending order in the table; add the frequencies from the top: $4+5 = 9 : 4+5+13 = 22$ $\therefore$ the $20\frac{1}{2}$th term lies in the class of thirteen 42's.
$\therefore$ the median is 42 grams.

With reference to the histogram, the median is the value of the vertical line which divides the histogram into 2 equal areas. (The median of a triangle bisects the area of the triangle.)
The distribution is bisected by the median and is further divided into four by the **quartiles**. The **lower** quartile bisects the lower half and the **upper** quartile divides the upper half.
E.g. In the list 1, 3, 4, 5, 6, 11, 15, 19, 20, 30, the lower quartile is 4 and the upper quartile is 19.
In the list 1, 3, 4, 6, 7, 8, 9, 10, the lower quartile falls between 3 and 4, so use $3\frac{1}{2}$; the upper quartile falls between 8 and 9, so use $8\frac{1}{2}$.

The distribution can be divided into **percentiles**, usually 10%, 20%, etc. The lower quartile is the 25th, the median the 50th and the upper quartile the 75th percentiles.

Cumulative frequency

When the classes are grouped, or the sum of the frequencies large, the median is found by constructing a **cumulative frequency distribution** as in table 5. This shows the lengths of 250 leaves measured in an experiment given to the nearest mm. The following table is formed from the frequency distribution by adding in each

length mm	frequency	cumulative frequency
30–34	15	15
35–39	30	45
40–44	52	97
45–49	75	172
50–54	60	232
55–59	18	250

Table 5

frequency in succession. Draw a smooth curve as a result of plotting the upper limits of the class range against the cumulative frequency. The limits of the class intervals are as shown because again the measurements are to the nearest mm. See figure 67.

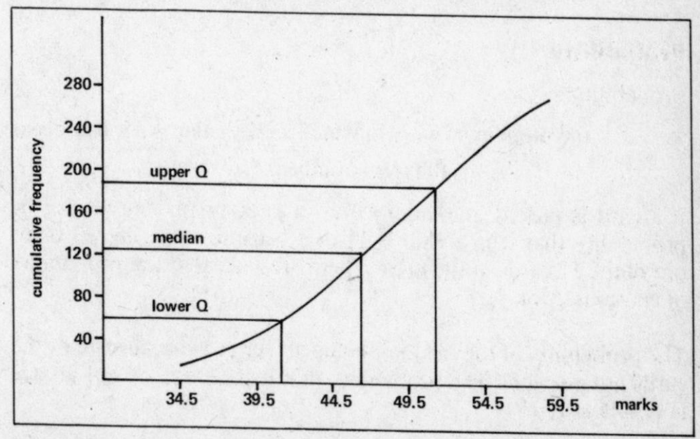

Figure 67

To estimate the median of a larger distribution, it is sufficient to find the number in the $\frac{1}{2}n$th position, i.e. the 125th number. Draw a horizontal line through the 125th leaf and read off the length of leaf given. The length of the median leaf is approximately 46 mm.

Similarly to estimate the lower quartile it is sufficient to locate the $\frac{1}{4}n$th term; the $\frac{3}{4}n$th term will locate the upper quartile. The lower quartile here is the 62·5th term, approximately 41·3 marks, the upper quartile is the 187·5th term about 50·6 marks.

There are several ways of measuring the spread across a frequency distribution. One is the **range**, which is the difference between the lowest and highest numbers of the distribution. The range in the table of egg weights (page 146) is $45-40=5$ grams. This does not give much detail of what is happening in the middle so use the **semi-interquartile range** which is $\frac{1}{2}$(upper quartile − lower quartile). In table 5 this will be $\frac{1}{2}(50{\cdot}6-41{\cdot}3)=\frac{1}{2}\times 9{\cdot}3=4{\cdot}65$.

The cumulative frequency curve can be used to find other information such as (i) what % of the leaves are greater than 50 mm? (ii) the 20% smallest leaves are to be treated: what is the largest leaf to be treated?

(i) Find the 50-mm mark on the length axis. Read off its cumulative frequency. It is 174. The first leaf which is greater than 50 mm is the 175th. The number of leaves greater than 50 mm = $250-174=76$ leaves. As a percentage this is $\frac{76}{250}\times 100=30{\cdot}4\%$.

(ii) 20% of 250 = 50. Find the length of the 50th leaf. This length = 40·3 mm. The longest leaf to be treated is therefore 40·3 mm.

Probability

Probability =

$$\frac{\text{the number of ways in which a particular event can occur}}{\text{the total number of all events}}$$

If a card is picked at random from a pack of playing cards, the probability that it is a club is $\frac{13}{52}$ or $\frac{1}{4}$ because there are 13 clubs out of the 52 cards in the pack. There are 4 aces, so the probability of an ace is $\frac{4}{52}$ or $\frac{1}{13}$.

The probability of the card not being an ace is $\frac{48}{52}$ because 48 of the cards are not aces. The probability that it is an ace or not an ace is $\frac{4}{52}+\frac{48}{52}=1$.

If p = the probability of an event occurring and q = the probability of an event not occurring then $p+q=1$ where p and q are fractions between 0 and 1. If the probability of an event occurring is $\frac{1}{3}$ then the probability of the non-event is $1-\frac{1}{3}=\frac{2}{3}$. Sometimes the probability is stated as a decimal or even as a percentage, so if the probability of an event occurring is 0·7 or 70% the probability of the non-event is 0·3 or 30%.

If a die is cast the probability of a 5 coming up is $\frac{1}{6}$. If the die is cast 300 times we would expect to see $\frac{1}{6}\times 300=50$ occasions when the 5 came up. This is called **expectation**.

Example Three coins are tossed together. What is the probability that (i) 3 heads come up? (ii) 2 are heads and one is tails? (iii) at least 1 is a tail?

Write down all the possibilities in a table:

1st coin	2nd coin	3rd coin
H	*H*	*H*
H	*H*	*T*
H	*T*	*H*
H	*T*	*T*
T	*T*	*H*
T	*T*	*T*
T	*H*	*H*
T	*H*	*T*

(i) There are 8 results, only 1 being all heads ∴ the required probability $=\frac{1}{8}$.
(ii) 2 heads and a tail occur 3 times ∴ the probability $=\frac{3}{8}$.
(iii) We require those results with 1, 2 or 3 tails. There are 7 of them, so the probability is $\frac{7}{8}$. Note that the probability of all heads is $\frac{1}{8}$; this means that there are no tails. Hence the probability of at least 1 tail coming up is $1-\frac{1}{8}=\frac{7}{8}$.

It is worth remembering that if 1 coin is tossed there are 2 possible events: head or tail. If 2 coins are tossed there are $2^2=4$ events. If 3 coins are tossed there are $2^3=8$ events, and n coins give 2^n events. Similarly, when 1 die is cast 6 events are possible, 2 dice give 6^2 or 36, 3 dice give 6^3 or 216 and n dice give 6^n events.

Example Two dice are thrown together. What is the chance (probability) that the total score is 8? The total can be achieved by scores of $6+2$, $5+3$, $2+6$, $3+5$ and $4+4$. Notice that $6+2$ and $2+6$ are different events.

There are 5 possible events $\therefore$ the chance is $\frac{5}{36}$.

shoe sizes	3	4	5	6	7	8	9
frequency	4	7	12	8	7	2	0

Table 6

Example Table 6 shows the shoe sizes of 40 people. If one person is taken at random what is the chance that his shoe size is (i) size 6? (ii) at least size 6?

(i) There are 8 people who take size 6, $\therefore$ the chance is $\frac{8}{40}$ or $\frac{1}{5}$.

(ii) At least size 6 covers the people taking sizes 6, 7, 8, 9. There are $8+7+2+0 = 17$ of them $\therefore$ the chance is $\frac{17}{40}$.

Total and compound probability

The chance of a 5 coming up when a die is cast is $\frac{1}{6}$.
The chance of a 6 coming up when a die is cast is $\frac{1}{6}$.
The chance of either a 5 or 6 coming up is $\frac{1}{6}+\frac{1}{6} = \frac{1}{3}$.
This is an example of **total probability**, when the events are mutually exclusive. One event has no effect on the other.

If two dice are thrown together the chance of the first being a 2 is $\frac{1}{6}$. The chance that the second is a 2 is also $\frac{1}{6}$. The chance that they are both 2's is $\frac{1}{6}\times\frac{1}{6} = \frac{1}{36}$.
This is an example of the law of **compound probability**, which states that if two independent events have probabilities p and q, the probability that they both take place is pq.

Example Two cards are picked at random from a pack of playing cards. What is the chance that (i) they are both aces? (ii) neither is an ace? (iii) one is a king and the other an ace?

(i) Treat this problem as one card drawn, followed by another, without returning the first.
The chance that the first is an ace is $\frac{4}{52}$. Assume success, so there are 3 aces left in the pack of 51 cards. $\therefore$ the chance of the second being an ace is $\frac{3}{51}$. The chance of them both being aces is $\frac{4}{52}\times\frac{3}{51} = \frac{1}{13}\times\frac{1}{17} = \frac{1}{221}$.

(ii) The chance that the first is not an ace is $1-\frac{4}{52} = \frac{48}{52} = \frac{12}{13}$.
Assume failure, so there are 4 aces left in the pack of 51 cards. The chance of the second not being an ace is $1-\frac{4}{51} = \frac{47}{51}$ $\therefore$ the chance of the two not being aces is $\frac{12}{13}\times\frac{47}{51}$.

(iii) The chance of the first being a king is $\frac{4}{52}$. The chance of the second being an ace is $\frac{4}{51}$. The compound probability is $\frac{4}{52}\times\frac{4}{51}$. But they could have been drawn in reverse order, doubling the chances: $\therefore$ the chance of an ace and a king is $2\times\frac{4}{52}\times\frac{4}{51} = \frac{8}{663}$.

Tree diagrams

A **tree diagram** is a systematic way of producing all the possibilities and can be used when problems seem to be more complicated.

Example An unusual die has the numbers 1, 1, 1, 2, 2, 3 on its six faces. It is thrown twice. Find the chance that (i) the same score is obtained with both throws; (ii) the first is greater than the second.

The chance of a 1 coming up is $\frac{3}{6}$ or $\frac{1}{2}$, of a two coming up is $\frac{2}{6}$ or $\frac{1}{3}$, of a 3 coming up is $\frac{1}{6}$. Figure 68(a) shows the tree diagram. From O there are 3 branches showing the first throw with its possible scores and their chances. Whatever the score of the first throw, there are three possibilities for the second throw. These are drawn in with their chances. Answer the questions by tracing along the branches.

(i) For the same scores trace along the 1 and 1 branches. Multiply the chances. Repeat for the 2 and 2, 3 and 3.
The chance of 1 and 1 $= \frac{1}{2} \times \frac{1}{2} = \frac{1}{4}$.
The chance of 2 and 2 $= \frac{1}{3} \times \frac{1}{3} = \frac{1}{9}$.
The chance of 3 and 3 $= \frac{1}{6} \times \frac{1}{6} = \frac{1}{36}$.
The total chance of the same scores $= \frac{1}{4} + \frac{1}{9} + \frac{1}{36} = \frac{14}{36}$ or $\frac{7}{18}$.

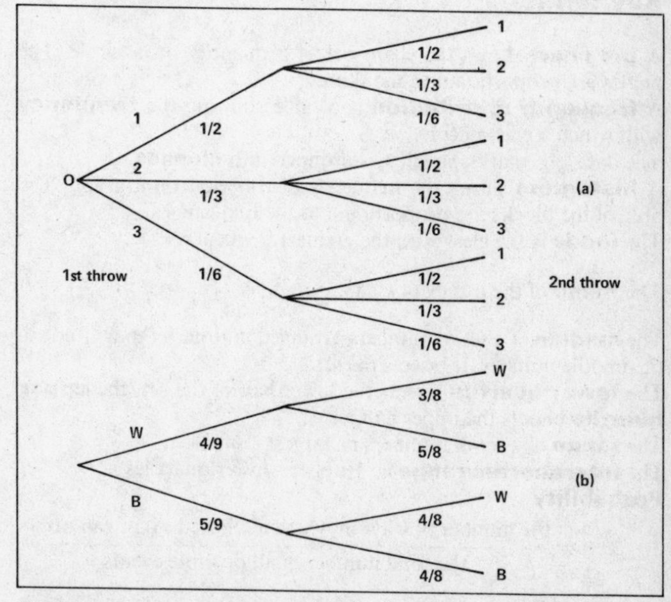

Figure 68

(ii) The first to be greater than the second:
2 on the first with 1 on the second has chance $= \frac{1}{3} \times \frac{1}{2} = \frac{1}{6}$;
3 on the first with 1 on the second has chance $= \frac{1}{6} \times \frac{1}{2} = \frac{1}{12}$;
3 on the first with 2 on the second has chance $= \frac{1}{6} \times \frac{1}{3} = \frac{1}{18}$.

The total chance is $\frac{1}{6} + \frac{1}{12} + \frac{1}{18} = \frac{6+3+2}{36} = \frac{11}{36}$.

Example A bag contains 4 white and 5 black balls. Two are drawn together. Find the chance that (i) both are white (ii) one is white and one is black.

Figure 68(b) shows the tree diagram. Assume the first ball drawn **is not replaced** so that the second ball is drawn from a total of eight. The chance of the first being white is $\frac{4}{9}$ and the other white is $\frac{3}{8}$. Repeat with the blacks.

(i) The chance that both are white is $\frac{4}{9} \times \frac{3}{8} = \frac{1}{6}$.
(ii) The chance of white and then black $= \frac{4}{9} \times \frac{5}{8} = \frac{5}{18}$. The chance of black and then white is $\frac{5}{9} \times \frac{4}{8} = \frac{5}{18}$; the total chance of a white and black $= 2 \times \frac{5}{18} = \frac{5}{9}$.

Key terms

A **pie chart** shows the share-out of a quantity on a circle. The angles are proportional to the shares.
A **frequency distribution** is a table showing the **frequency** with which a class occurs.
The data, e.g. marks, weights, is grouped into **classes**.
A **histogram** shows the frequency distribution on a graph. The area of the blocks are proportional to the frequencies.
The **mode** is the class with the greatest frequency.

The **mean** of the numbers $x_1, x_2, x_3, \ldots x_n$ is $\frac{1}{n}(x_1 + x_2 + \cdots x_n)$.

The **median** of a set of numbers arranged in order of magnitude is the middle number. It bisects the set.
The **lower quartile** bisects the lower half of the set, the **upper quartile** bisects the upper half.
The **range** of a set of numbers is: largest − smallest.
The **interquartile range** is: $\frac{1}{2}$(upper − lower quartiles).
Probability

$$= \frac{\text{the number of ways in which a selected event can occur}}{\text{the total number of all possible events}}$$

Chapter 14
Trigonometry (1)

In figure 69(a) the names of the sides refer to angle A. Pairing them off as fractions produces the trigonometrical ratios:

$$\text{sine } A = \frac{\text{opposite}}{\text{hypotenuse}} \quad \cos A = \frac{\text{adjacent}}{\text{hypotenuse}} \quad \tan A = \frac{\text{opposite}}{\text{adjacent}}$$

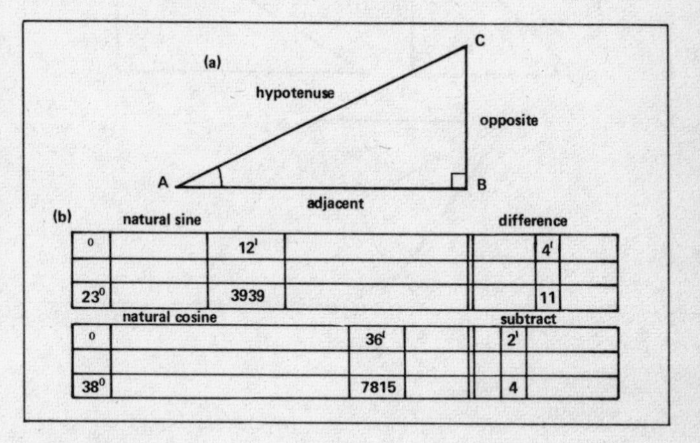

Figure 69

Use of tables

The 4-figure tables give the ratios as 4-figure decimals. Figure 69(b) shows the relevant numbers required to find sin 23°16′. On line 23° column 12′ (the nearest below 16′) we find 0·3939. The difference is found in column 4′ on the same line, 11.

Sin 23°16′ = 0·3939 + 0·0011 = 0·3950.

In reverse, to find the angle whose cosine is 0·7819. Figure 69(b) gives the relevant numbers from the natural cosine table. Look for the nearest number below 0·7819. It is 0·7815 on line 38, column 36, the difference of 4 is found in the difference column—column 2′. The required angle = 38°36′ − 2′ = 38°34′.

Always look for the number or angle below that required. **Add** the differences with sine and tangent, **subtract** with cosine.

3-figure tables, using decimal part of degrees, have no differences.

Sin 23·2° = 0·394 at the intersection of line 23° and column ·2°.

In figure 70(a) use tables to find (i) x and (ii) angle B.

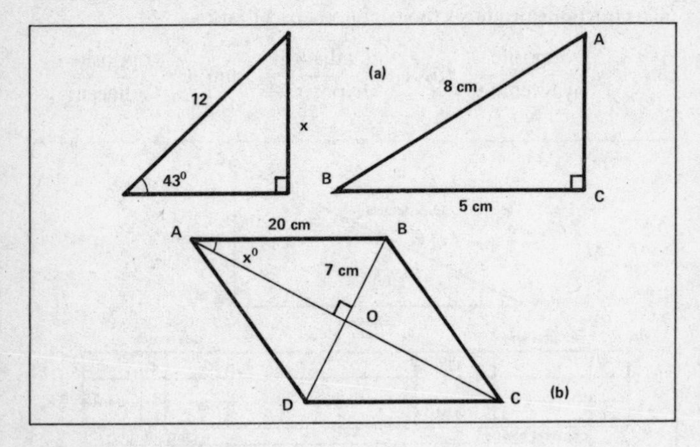

Figure 70

(i) $\dfrac{\text{opposite}}{\text{hypotenuse}} = \dfrac{x}{12} = \sin 43°, \dfrac{x}{12} = 0{\cdot}6820 \Leftrightarrow x = 12 \times 0{\cdot}6820$

$x = 8{\cdot}18$ cm to 3 significant figures.

Determine the ratio, find sin 43° in the tables and solve the resulting equation, multiplying both sides by 12.

(ii) $\dfrac{\text{adjacent}}{\text{hypotenuse}} = \dfrac{5}{8} = 0{\cdot}625 = \cos B$

angle $B = 51°24' - 5' = 51°19'$

Determine the ratio, write it as a 4-figure decimal, then find the angle with this number in the cosine table as shown.

Example *ABCD* is a rhombus of side 20 cm. The shorter diagonal is 14 cm. Calculate (i) the angles of *ABCD*; (ii) the longer diagonal.

(i) In figure 70(b) we must know that the diagonals of a rhombus bisect at right angles. So $OB = 7$ cm (half diagonal). In $\triangle OBA$

$$\frac{\text{opposite}}{\text{hypotenuse}} = \frac{7}{20} = 0{\cdot}3500 = \sin x^\circ \quad \therefore \quad x^\circ = 20^\circ 29'.$$

$\therefore \quad A = 20^\circ 29' \times 2 = 40^\circ 58'$ and

$B = 180^\circ - 40^\circ 58' = 139^\circ 2'$

(ii) Using Pythagoras' theorem $AO^2 = 20^2 - 7^2 \; \therefore \; AO = \sqrt{351} =$ 18·73, so the longer diagonal = 37·5 cm to 3 significant figures.

Elevation and depression

Figure 71(a) shows the angle of **elevation** of C from A, the 'raised' angle, and the angle of **depression** of A from C, the 'lowered' angle. They are equal to each other (alternate angles).

Bearings

A **bearing** is an angle between 000° and 360° (always give 3 figures). In figure 71(b) to find the bearing of B from A, stand at A,

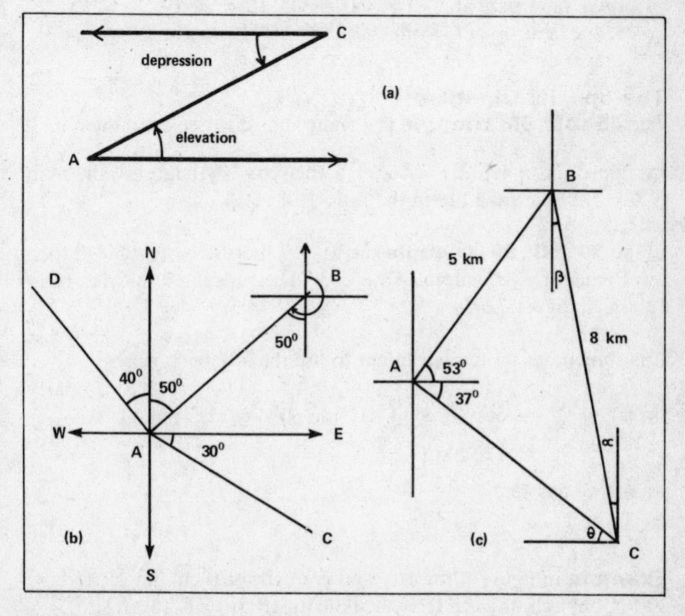

Figure 71

look north, turn clockwise through 50° to face B. The bearing is 050°. The bearings of C, D from A are 120° and 320°. We still use the 4 cardinal points N, S, E, W and the points NE, SE, SW, NW. Note that E is 090°, S is 180°, W is 270°. NE, SE, SW, NW are 045°, 135°, 225° and 315° respectively.

The bearing of A from B is found by standing at B, turning clockwise to face A. The bearing $= 180° + 50° = 230°$. Verify that the bearing of A from C is 300° and that of A from D is 140°.

Example B and C are 8 km apart. B is 5 km from A on a bearing 037° and C is on a bearing 127° from A. Calculate (i) the bearing of B from C; (ii) the bearing of C from B.

Figure 71(c) shows that angle $BAC = 53° + 37° = 90°$.

(i) Find angle BCA. $\frac{5}{8} = \sin BCA = 0{\cdot}6250$ $\therefore$ angle $BCA = 38°41'$. Draw N, S, E, W lines through C. $\theta = 37°$ (alternate angles) $\therefore$ the bearing of B from C is $270° + 37° + 38°41' = 345°41'$.

(ii) Draw N, S, E and W lines through B.
$\alpha = \beta$ (alternate angles) $= 360° - 345°41' = 14°19'$
$\therefore$ the bearing of C from B is $180° - 14°19' = 165°41'$.

The special triangles

The **45°, 45°, 90° triangle** is a right-angled isosceles triangle.

In figure 72(a) let $XY = YZ = 1$ then by Pythagoras' theorem $XZ = \sqrt{2}$. The sides are in the ratio $1 : 1 : \sqrt{2}$.

In the **30°, 60°, 90° triangle** $\sin 30° = \frac{1}{2}$. Let $BC = 1$, $AC = 2$ then by Pythagoras' theorem $AB = \sqrt{3}$. The sides are in the ratio $1 : \sqrt{3} : 2$ (figure 72(b)).

It is sometimes more convenient to use the following ratios:

$$\sin 60° = \frac{\sqrt{3}}{2}, \quad \cos 30° = \frac{\sqrt{3}}{2}, \quad \tan 60° = \sqrt{3}, \quad \tan 30° = \frac{1}{\sqrt{3}},$$

$$\sin 45° = \cos 45° = \frac{1}{\sqrt{2}}$$

Example In figure 72(c) $AC = 50$ cm, $AD = 30$ cm, angle $ABE = 30°$. XA bisects angle EAB. Calculate (i) AB; (ii) AE; (iii) EX.

(i) $AD = BC = 30$ cm. In $\triangle ABC$ $AB = 40$ cm (a $3 : 4 : 5$ triangle).

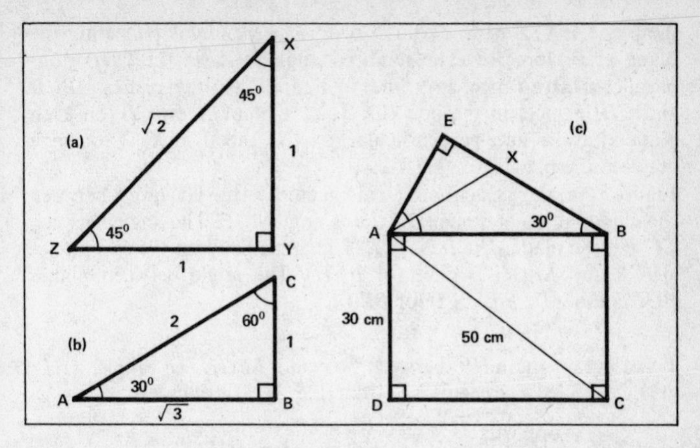

Figure 72

(ii) $\triangle ABE$ is a $30°, 60°, 90°$ triangle. $\therefore$ $AE = 20\,\text{cm}$ ($\frac{1}{2}AB$).

(iii) In $\triangle AEX$ $\dfrac{\text{opposite}}{\text{adjacent}} = \dfrac{EX}{20} = \tan 30°$ $\therefore$ $\dfrac{EX}{20} = 0{\cdot}577\,4 \Leftrightarrow$

$$EX = 20 \times 0{\cdot}5774 \quad \therefore \quad EX = 11{\cdot}55 \text{ cm.}$$

Three-dimensional problems

The following concepts should be appreciated: (1) The angle between a **line** and a **plane**. In figure 73(a) AB is a line meeting

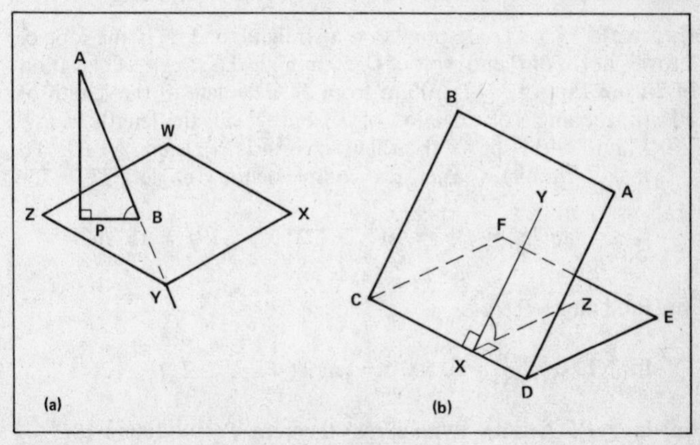

Figure 73

the plane *WXYZ* at *B*. From *A* drop a perpendicular meeting the plane at *P*. Join *PB*. The required angle is *ABP*. (2) **Two** non-parallel **planes** meet in a line. In figure 73(b) the planes *ABCD* and *CDEF* meet in the line *CD*. Take a point *X* on *CD*, on each plane draw a line perpendicular to *CD* through *X*. The angle between the planes is angle *YXZ*.
Figure 74(a) shows a cuboid 8 cm × 6 cm × 7 cm. The angle between the diagonal *AG* and plane *EFGH* is angle *AGE*. The angle between *EC* and the plane *ABCD* is *ACE*. The angle between the two planes *ABGF* and *EFGH* is *AFE* (or *BGH*). The angle between planes *BCFE* and *ADFE* is *CFD* (or *BEA*).

To calculate the angle between *AG* and *EFGH*, i.e. angle *AGE*: in right-angled $\triangle EFG$ $EG = 10$ cm (a 6 : 8 : 10 triangle);

$$\text{in } \triangle AEG \quad \frac{\text{opposite}}{\text{adjacent}} = \frac{7}{10} = 0{\cdot}7000 = \tan AGE$$

$\therefore$ angle $AGE = 35°$.

To calculate the angle between the planes *BCFE* and *ADFE*, i.e. angle *CFD*:

$$\text{In } \triangle CFD \quad \frac{\text{opposite}}{\text{adjacent}} = \frac{8}{7} = 1{\cdot}1429 = \tan CFD$$

$\therefore$ angle $CFD = 48°49'$.

Example *P* and *Q* are points on a straight road. *F* is the foot of a tower north of *P* and west of *Q*, 50 m high. The angle of elevation of *T* from *P* is 15°. *Q* is 100 m from *F*. Calculate (i) the length of *PF*; (ii) the angle of elevation of *T* from *Q*; (iii) the length of *PQ*.

(i) Figure 74(b) shows the diagram. Find *PF* from $\triangle FPT$. To make arithmetic easier, use complementary angle $PTF = 75°$

$$\frac{PF}{50} = \tan 75° \Leftrightarrow PF = 50 \times 3{\cdot}7321 \quad \therefore \quad PF = 187\,\text{m}.$$

(ii) Find angle *TQF*.

$$\text{In } \triangle TFQ \quad \frac{50}{100} = 0{\cdot}5000 = \tan TQF \quad \therefore \quad TQF = 26°34'.$$

(iii) In $\triangle PFQ$ using Pythagoras' theorem $PQ^2 = 100^2 + 187^2 = 10\,000 + 34\,970$. $PQ = \sqrt{44\,970}$ $\therefore$ $PQ = 212$ m.

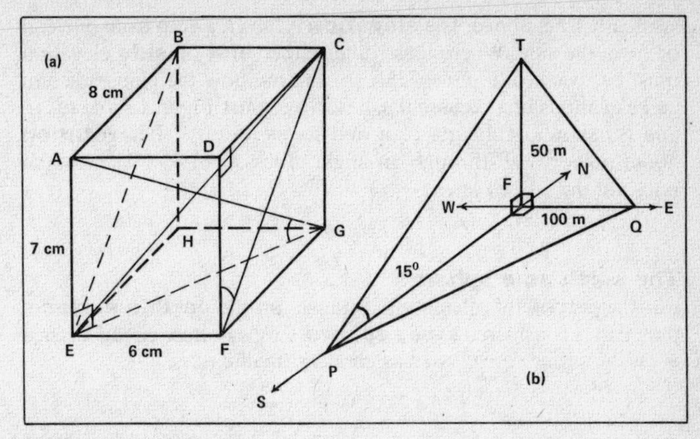

Figure 74

Plan and elevation

Hĕre, the meaning of the terms rather than the detail of construction will be considered. The **plan** of an object is the view seen

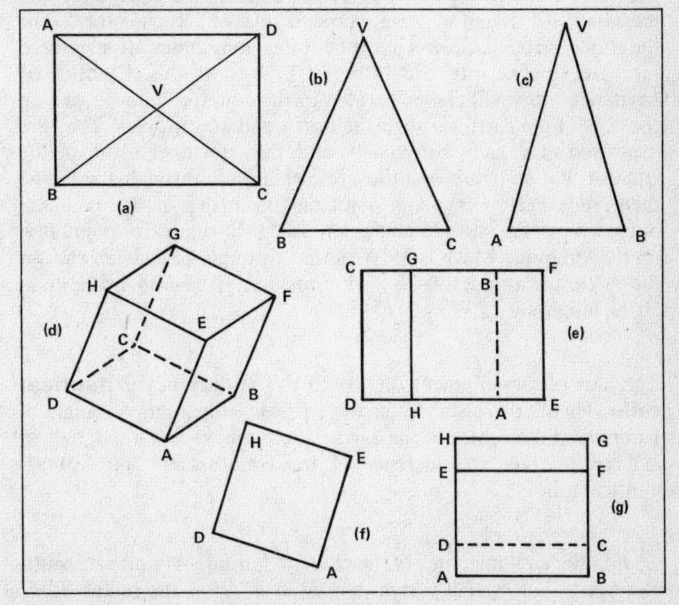

Figure 75

vertically from above. The **elevation** is the view seen from one end or from the side. Where these differ either **end** or **side** elevation must be specified. Figures 75(a), (b) and (c) show the plan, side and end elevations of a rectangular-based pyramid. Figures 75(d), (e), (f) and (g) show a cube, its plan and its elevations, after **rotation** about an edge AB through an angle of less than 45°. The broken lines are the unseen lines.

The earth as a sphere
For the purpose of calculating distances on the **earth's surface**, the earth is considered to be a **sphere**. Circles whose circumference is on the surface are classed as great or small circles.

Great circles have their centre at the centre of the earth. Their radius is the radius of the earth. **Small circles** have a radius smaller than the earth's radius. The equator is a great circle, a **meridian of longitude** is half of a great circle passing through the north and south poles, e.g. NPS in figure 76. The equator and meridians of longitude are perpendicular to each other. The longitude through Greenwich is 0°; other longitudes are measured in degrees between 0° and 180° east or west of this. **Circles of latitude** are small circles parallel to the equator, with centres on the axis of the earth through the north and south poles. They are measured in degrees between 0° and 180° north or south of the equator. E.g. In figure 76 if the circle of latitude through P is 50°N, then angle $POQ = 50°$. Any point on the earth's surface is at the intersection of a latitude and a longitude. In figure 76, point P is on the longitude which is 35°W of that through Greenwich and on the circle of latitude 50°N, so its position is said to be latitude 50°N, longitude 35°W.

The unit of measurement on the earth's surface is the **nautical mile**. This is the distance on the equator subtending an angle of 1 minute at the centre of the earth. There are 60′ in 1° and 360° in a circle $\therefore$ the circumference of the equator $60 \times 360 = 21\,600$ nautical miles.

To find the circumference of the circle of latitude $A°$ north or south (see figure 76): $\triangle OPB$ is right-angled at B. OP is the radius 'R' of the earth, BP is the radius 'r' of the circle of latitude $A°$ north, BP

is parallel to OQ $\therefore$ $BPO = A°$ (alternate angles).

Using trigonometry: in $\triangle OPB$, $\frac{r}{R} = \cos A°$ $\therefore$ $r = R \cos A°$ and so the circumference of the circle of latitude $= 2\pi R \cos A°$.

But $2\pi R$ = circumference of the earth.

$\therefore$ Circumference of circle of latitude $A°$ north = Circumference of earth $\times \cos A°$.

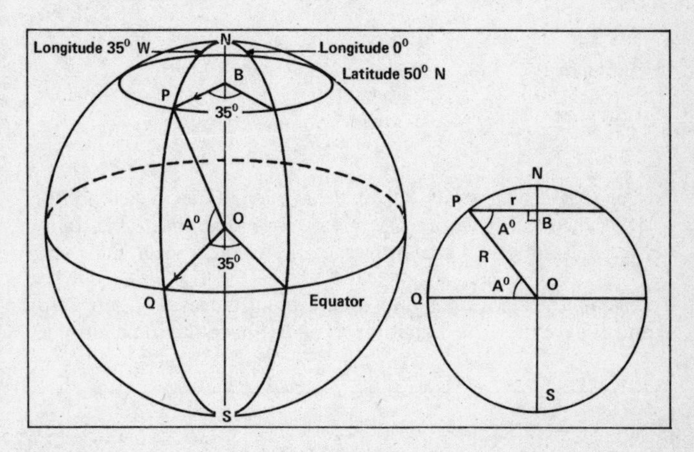

Figure 76

To calculate the distances on great and small circles given the positions of points: in figure 77(a) R and T are on the equator and have the same longitudes as P and Q respectively, P is at latitude 63°N, longitude 45°W, and Q is at latitude 63°N, longitude 35°E.

To calculate the distance RT, on the equator, a great circle.

Angle $ROT = 35° + 45° = 80°$. RT subtends an angle of 80° at the centre of the earth.

$$\therefore \quad \frac{\text{arc } RT}{\text{circumference}} = \frac{\text{angle } ROT}{360°} \Leftrightarrow \frac{\text{arc } RT}{21\,600} = \frac{80}{360} \Leftrightarrow$$

$$\text{arc } RT = 21\,600 \times \frac{80}{360} = 60 \times 80 = 4\,800 \text{ nm (nautical miles).}$$

To calculate the distance PQ which is the arc of a small circle:

Circumference of circle of latitude 63°N = Circumference of earth × cos 63°. ∴ To find arc PQ, first find arc RT and multiply this by cos 63°. RT = 4 800 nm ∴ PQ = 4 800 × cos 63° = 4 800 × 0·454 0 = 2 180 nm.

To calculate the distance RP which is an arc of a meridian of longitude, i.e. the arc of a great circle. P is on latitude 63°N and R is on the equator, ∴ angle POR = 63°.

$$\frac{\text{arc } PR}{\text{circumference}} = \frac{63}{360} \Leftrightarrow \frac{\text{arc } PR}{21\,600} = \frac{7}{40} \Leftrightarrow$$

$$\text{arc } PR = \frac{7}{40} \times 21\,600 = 7 \times 540 \quad \therefore \quad \text{arc } PR = 3\,780 \text{ nm}$$

Notice that if two points A and B have longitudes which add up to 180°, then AB is an arc of a great circle. The shortest distance between A and B, on the surface, is the distance over the poles.

To calculate the positions of points given distances: in figure 77(b) A and B have the same longitude, B and C have the same latitude.

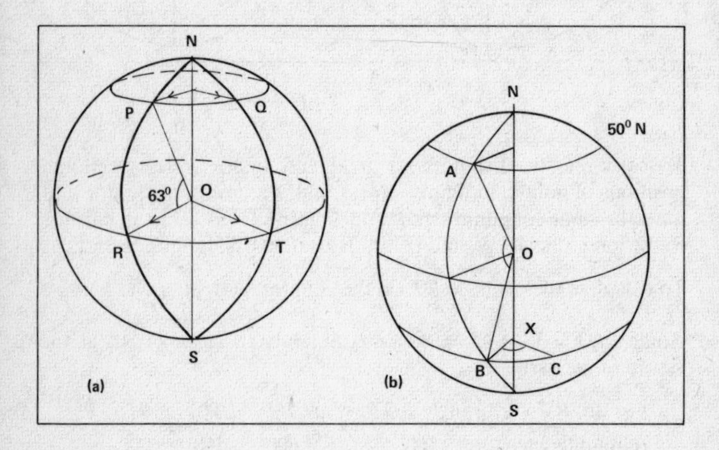

Figure 77

A is latitude 50°N, longitude 43°W. Arc AB = 6 600 nm, arc BC = 2 490 nm.

To calculate the position of B, we require angle AOB. Arc AB is an arc of a great circle.

$$\frac{\text{arc } AB}{21\,600} = \frac{\text{angle } AOB}{360} \quad \therefore \quad \frac{6\,600}{21\,600} = \frac{\text{angle } AOB}{360} \Leftrightarrow$$

$$\therefore \quad \text{angle } AOB = \frac{6\,600}{21\,600} \times 360 = 110^\circ$$

$\therefore$ B is 110° south of A, $\therefore$ B is on latitude $110-50 = 60^\circ$S and the position of B is latitude 60°S, longitude 43°W.

To calculate the position of C, we require angle BXC. BC is an arc of a small circle latitude 60°S.

Circumference of this circle $= 21\,600 \cos 60^\circ = 21\,600 \times \frac{1}{2} = 10\,800$

$$\frac{2490}{10\,800} = \frac{\text{angle } BXC}{360} \Leftrightarrow \text{angle } BXC = \frac{2490}{10\,800} \times 360^\circ$$

$$\therefore \quad \text{angle } BXC = 83^\circ.$$

$\therefore$ C is 83° to the east of B, so it is on longitude $83-43 = 40^\circ$E. The position of C is latitude 60°S, longitude 40°E.

Key terms

Sine, **cosine** and **tangent** are three of the trigonometric functions. The graphs of $f(x) = \sin x^\circ, \cos x^\circ$, and $\tan x^\circ$ are shown on page 171, the definitions given on page 168. Sines, cosines and tangents are used to find sides and angles in triangles, using right-angled-triangle trigonometry with the modified definitions stated on page 157, or using the sine or cosine rules as described on page 174.
The angles of **elevation and depression** are the names given to the two angles measured upwards and downwards from the horizontal. (See the diagram in figure 71.)
To state the angular position of a point B from a point A a **bearing** (the angle described by facing north and turning clockwise to face B when standing at A) is given.
The **plan and elevation** of an object are the views seen from above and from the side or end respectively.
A **great circle** has a radius equal to the earth's radius.
A **small circle** has a radius less than the earth's radius.
A **circle of latitude** is a small circle parallel to the equator.
A **meridian of longitude** is half of a great circle through the north and south poles.

Chapter 15
Trigonometry (2)

In figure 78(a) assuming the positions of the x and y axes on a graph, 0° is on the $+x$ axis and **positive** angles are measured **anti-clockwise**. 360° will also be on the $+x$ axis. Each quarter of the circle is called a **quadrant**. To determine the sine, cosine and tangent of any angle between 0° and 360° we modify the definition given on page 157. Rotate the radius r to OP_1 through the required angle $A°$ and define the ratios as follows:

$$\sin A = \frac{y \text{ co-ordinate of } P_1}{\text{radius}} \qquad \cos A = \frac{x \text{ co-ordinate of } P_1}{\text{radius}}$$

$$\tan A = \frac{y \text{ co-ordinate of } P_1}{x \text{ co-ordinate of } P_1}$$ The radius is assumed positive.

1st quadrant: if $A = 20°$ the x and y co-ordinates of P are both positive (+). Use $\triangle P_1OM$.
2nd quadrant: if $A = 160°$, angle $P_2ON = 180° - 160° = 20°$. $\triangle$s P_2ON and P_1OM are congruent. The x co-ordinate of P_2 is negative (−), the y is positive (+).
3rd quadrant: if $A = 200°$, angle $P_3ON = 200° - 180° = 20°$. $\triangle$s P_3ON and P_1OM are congruent. The x and y co-ordinates of P are both negative (−).
4th quadrant: if $A = 340°$, angle $P_4OM = 360° - 340° = 20°$. $\triangle$s P_4OM and P_1OM are congruent. The y co-ordinate of P is negative (−), the x positive (+).
Using these values of A the ratios appear as follows:

1st quadrant:

$$\sin 20° = \frac{+y}{r} = +0{\cdot}3420$$

$$\cos 20° = \frac{+x}{r} = +0{\cdot}9397$$

$$\tan 20° = \frac{+y}{+x} = +0{\cdot}3640$$

2nd quadrant:

$$\sin 160° = \frac{+y}{r} = +0{\cdot}3420$$

$$\cos 160° = \frac{-x}{r} = -0{\cdot}9397$$

$$\tan 160° = \frac{+y}{-x} = -0{\cdot}3640$$

3rd quadrant:

$$\sin 200° = \frac{-y}{r} = -0{\cdot}3420$$

4th quadrant:

$$\sin 340° = \frac{-y}{r} = -0{\cdot}3420$$

$$\cos 200^\circ = \frac{-x}{r} = -0{\cdot}9397 \qquad \cos 340^\circ = \frac{+x}{r} = +0{\cdot}9397$$

$$\tan 200^\circ = \frac{-y}{-x} = +0{\cdot}3640 \qquad \tan 340^\circ = \frac{-y}{+x} = -0{\cdot}3640$$

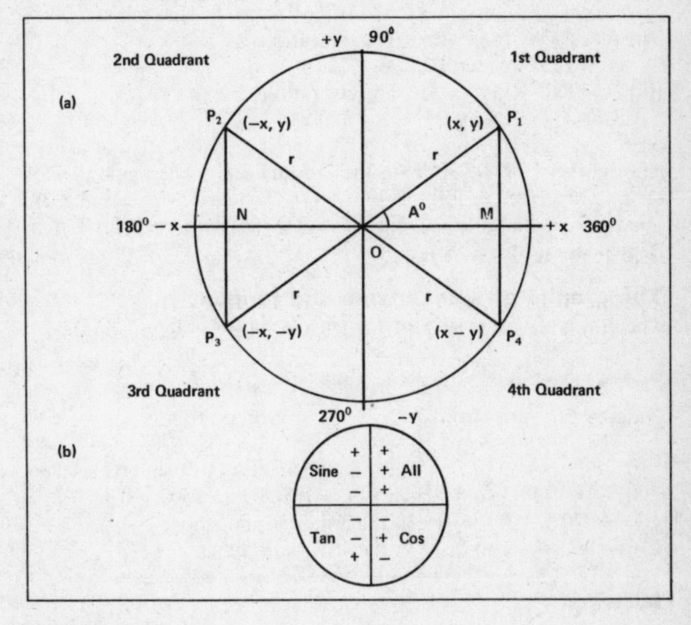

Figure 78

In general, we can see from this example that we need to know: (i) the angle to look up in the tables; (ii) the sign of the ratio.

If A° in 1st quadrant (i) look up angle A° (ii) **A**ll the ratios are +.
If A° in 2nd quadrant (i) look up $180^\circ - A^\circ$ (ii) **S**ine only is +.
If A° in 3rd quadrant (i) look up $A^\circ - 180^\circ$ (ii) **T**an only is +.
If A° in 4th quadrant (i) look up $360^\circ - A^\circ$ (ii) **C**os only is +.

Figure 78(b) shows a summary of signs of the ratios. The initials of the + ratios can be remembered by the phrase "**A**ll **S**tations **T**o **C**rewe" which has initials ASTC.

When finding the value of sin 230°, etc., it is advisable to draw a diagram each time and work out the details from it, keeping the above generalisations in mind.

Example Find (i) sin 230°; (ii) tan 143°; (iii) cos 337°.

(i) In figure 79 $a = 230° - 180° = 50°$. The sine ratio is $\mp$
$\therefore$ sin 230° = − sin 50° = −0·766 0 from the sine tables.

(ii) $b = 180° - 143° = 37°$. The tan ratio is $\mp$.
Tan 143° = − tan 37° = −0·753 6.

(iii) $c = 360° - 337° = 23°$. The cos ratio is $\pm$.
Cos 337° = + cos 23° = +0·920 5.

Alternatively for (i) 230° is in the 3rd quadrant. The angle to look up is 230° − 180° = 150°. From ASTC the sine is − in the 3rd quadrant $\therefore$ sin 230° = − sin 50° = −0·766 0. (ii) and (iii) can be treated similarly.

The graphs of sine, cosine and tangent

The graph of $f: x \to \sin x$ or $y = \sin x$ is shown in figure 80(a).

x increases:	sin x varies from:	Observations
0° − 90°	0 to 1	sin 0° = 0; sin 30° = 0·5; sin 60° = 0·866
90° − 180°	1 to 0	sin 120° = 0·866 sin 150° = 0·5
180° − 270°	0 to −1	In the 3rd quadrant sin x is −
270° − 360°	−1 to 0	In the 4th quadrant sin x is −

Table 7

Study table 7 to help you understand the shape of the curve.

Figure 80(a) also shows the graph of $f: x \to \cos x$ or $y = \cos x$. Cosine means 'the complement of the sine'. Complementary angles add up to 90° $\therefore \cos x = \sin(90° - x)$. The cosine curve is the sine curve translated 90° along the x axis in the negative direction.

The sine and cosine curves repeat every 360°. They are called **periodic** curves, the period being 360°. We have one **oscillation** between 0° and 360°. The greatest height, 1 unit, is the **amplitude**.

Figure 80(b) shows the graph of $f: x \to \tan x$ or $y = \tan x$. Tan 0° = 0, tan 45° = 1 and tan 90° = ∞. A vertical line is drawn through $x = 90°$ and the curve approaching it is shown. With the tangent

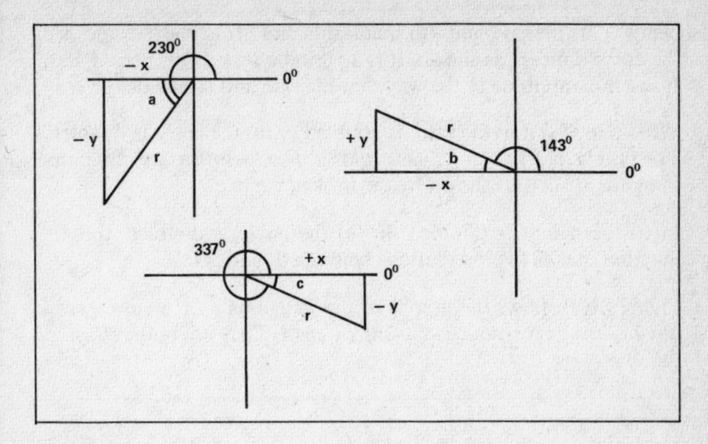

Figure 79

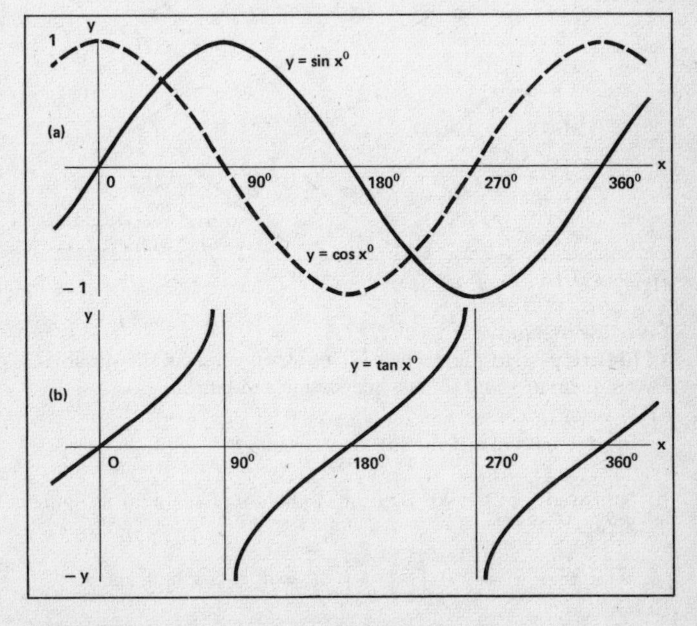

Figure 80

being $-$ in the 2nd and 4th quadrants and $+$ in the 1st and 3rd, the curve appears as shown. It is a periodic curve with period 180°. It has no amplitude in the way that the sine and cosine do.

What happens if (i) the ratio is doubled or (ii) the angle is doubled? The graphs of (i) $f: x \rightarrow 2 \sin x$ and (ii) $f: x \rightarrow \sin 2x$ are compared in figure 81(a). (i) is shown by the broken curve.

In (i) the height is doubled. In (ii) the angle is doubled and the function makes two oscillations between 0° and 360°.

Figure 81(b) shows the graphs of the functions $f: x \rightarrow \sin x - \cos x$ (the broken curve) and $f: x \rightarrow \sin x + \cos x$. They are both a form of the sine curve.

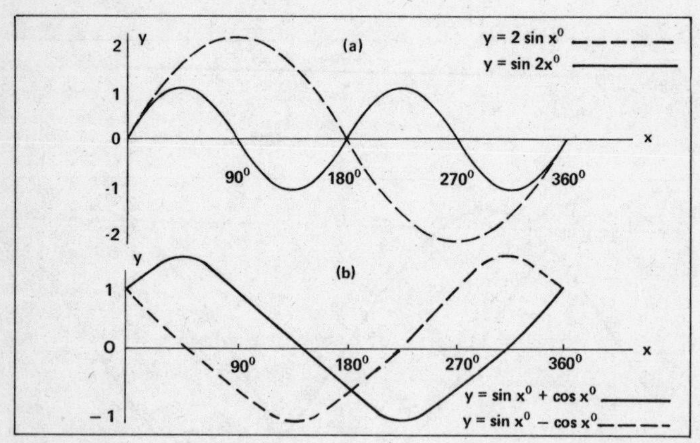

Figure 81

Two identities

An **identity** is a formula which is true for all values of the variables. Two of them are used to link sine, cosine and tangent.

(i) $\sin^2 A + \cos^2 A = 1$ (ii) $\dfrac{\sin A}{\cos A} = \tan A$

(i) is proved for $A < 90°$ by using Pythagoras' theorem in figure 82(a).

$$r^2 = x^2 + y^2 \Leftrightarrow 1 = \left(\frac{x}{r}\right)^2 + \left(\frac{y}{r}\right)^2 \Leftrightarrow 1 = \cos^2 A + \sin^2 A.$$

Notice that $\cos^2 A = (\cos A)^2$, the square of the ratio.

(ii) In figure 82(a) $\dfrac{\sin A}{\cos A} = \dfrac{y}{r} \div \dfrac{x}{r} = \dfrac{y}{r} \times \dfrac{r}{x} = \dfrac{y}{x} = \tan A$

Although shown only for $A < 90^\circ$ these identities can be proved true for any value of A.

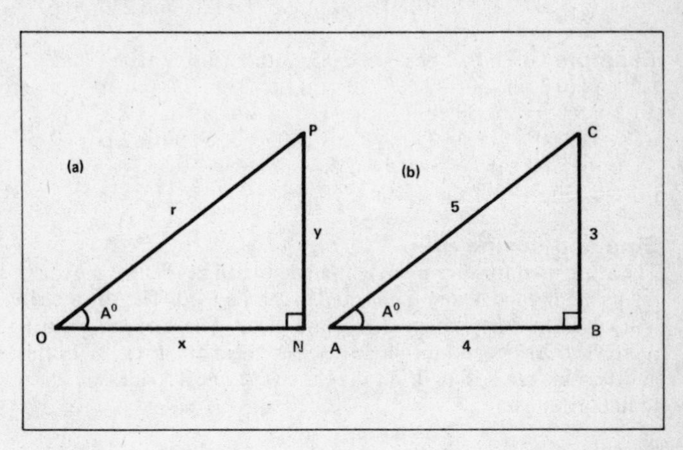

Figure 82

Example If A is acute and $\sin A = \frac{3}{5}$, find $\cos A$ and $\tan A$ without using tables.
Sketch right-angled $\triangle ABC$ as in figure 82(b), with $BC = 3$ (opposite) and $AC = 5$ (hypotenuse). $AB = 4$ (a 3,4,5 triangle).
$\therefore \cos A = \frac{4}{5}$ and $\tan A = \frac{3}{4}$ both ratios being +.

If A is given as an obtuse angle $\sin A = +\frac{3}{5}$ still and we can draw the triangle as above giving $\cos A = -\frac{4}{5}$ and $\tan A = -\frac{3}{4}$, remembering that cos and tan are negative (−) in the 2nd quadrant. Alternatively identity (i) can be applied:

$\cos^2 A + \sin^2 A = 1 \quad \therefore \quad \cos^2 A + \frac{9}{25} = 1 \Leftrightarrow \cos^2 A = 1 - \frac{9}{25} = \frac{16}{25}$

$\Leftrightarrow \cos A = \pm\frac{4}{5}$ and we choose the − sign for the obtuse angle. For the value of $\tan A$ use identity (ii)

$$\tan A = \frac{\sin A}{\cos A} = \frac{3}{5} \div \frac{-4}{5} = \frac{3}{5} \times \frac{5}{-4} = \frac{3}{-4} \qquad \tan A = -\frac{3}{4}$$

Example To find the values of x between 0° and 360° for which (i) $\sin x = 0{\cdot}8480$ and (ii) $\cos x = -\frac{1}{2}$.

(i) From the tables $x = 58°$. This is a positive ratio, $\therefore$ x can be a 2nd quadrant angle. $x = 180° - 58° = 122°$ as well. $\therefore x = 58°$ and $122°$.

(ii) The cosine is negative in the 2nd and 3rd quadrants. The angle of which the cosine is $+\frac{1}{2}$ is $60°$,
$\therefore x = 180° - 60°$ and $180° + 60°$, i.e. $x = 120°$ and $240°$.

Example Given that $\cos 60° = 0{\cdot}5$, without using tables find the value of (i) $3 \sin 150° - 2 \cos 240°$; (ii) $\sin^2 330°$.

(i) $\cos 60° = \sin 30° = 0{\cdot}5$: $\sin 150° = \sin 30°$ and $\cos 240° = -\cos 60°$ $\therefore$ $3 \sin 150° - 2 \cos 240° = 3 \sin 30° + 2 \cos 60° = 3 \times 0{\cdot}5 + 2 \times 0{\cdot}5 = 5 \times 0{\cdot}5 = 2{\cdot}5$.

(ii) $\sin 330° = -\sin 30°$ $\therefore \sin^2 330° = (-0{\cdot}5)^2 = +0{\cdot}25$.

Sine and cosine rule

These are used to solve non-right-angled triangles. Three measurements are needed before the formulae can be used. The **sine rule** is used when 2 angles and 1 side are given. The **cosine rule** is used when 2 sides and the angle between them are given. With this another side can be found. We use the cosine rule if 3 sides are given to find an angle.

This information coincides with the tests for congruent triangles stated on page 82.

Referring to $\triangle ABC$ in figure 83, the sine rule states

$$\frac{a}{\sin A} = \frac{b}{\sin B} = \frac{c}{\sin C}$$

The cosine rule is (i) $a^2 = b^2 + c^2 - 2bc \cos A$; (ii) $b^2 = c^2 + a^2 - 2ac \cos B$; (iii) $c^2 = a^2 + b^2 - 2ab \cos C$.

Example In $\triangle ABC$ angle $A = 110°$, angle $B = 20°$ and $b = 7$ cm. Find a.

2 angles and 1 side are given. Use the relevant part of the sine rule:

$$\frac{a}{\sin A} = \frac{b}{\sin B} \quad \therefore \quad \frac{a}{\sin 110°} = \frac{7}{\sin 20°}$$

$$\Leftrightarrow a = \frac{7 \sin 110°}{\sin 20°} = \frac{7 \sin 70°}{\sin 20°}$$

$a = 19{\cdot}2$ cm to 3 significant figures.

no.	log
7 sin 70°	0·8451 $\bar{1}$·9730
 sin 20°	0·8181 $\bar{1}$·5341
1923	1·2840

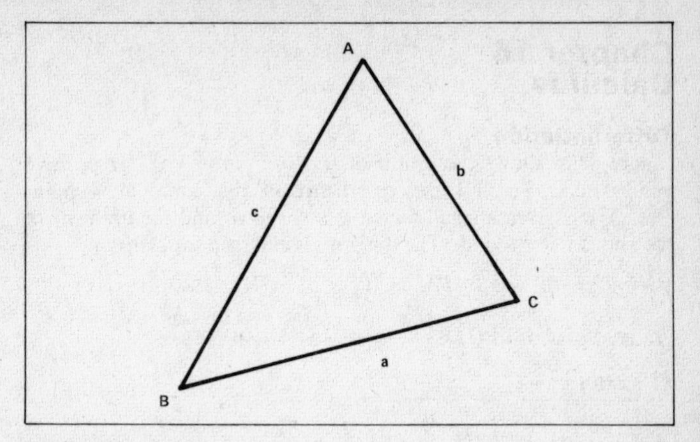

Figure 83

Notice that sin 110° = sin 70°. Logarithms are used just to show an example of such working. The logs of sin 70° and sin 20° can be found by looking straight into the logarithm sine table. A calculator gives sine 110° directly.

Example In $\triangle ABC$ $c = 4$ cm, $a = 10$ cm, angle $B = 52°$. Find b.

2 sides and the angle between them are given. Use cosine rule (ii).

$b^2 = 4^2 + 10^2 - 2 \times 4 \times 10 \cos 52° = 16 + 100 - 80 \cos 52°$

$b^2 = 116 - 80 \times 0{\cdot}6157 = 116 - 49{\cdot}26 = 66{\cdot}74 \Leftrightarrow b = 8{\cdot}17$ cm

Notice that $80 \times 0{\cdot}6157$ is worked out first, then the answer is subtracted from 116. Multiplication comes before subtraction.

Example In $\triangle ABC$ $a = 10$ cm, $b = 7$ cm, $c = 5$ cm. Find angle A.

Given three sides, use cosine rule (i).

$10^2 = 7^2 + 5^2 - 2 \times 7 \times 5 \cos A \Leftrightarrow 100 = 49 + 25 - 70 \cos A$

$\Leftrightarrow 100 - 49 - 25 = -70 \cos A \Leftrightarrow 26 = -70 \cos A$

$\Leftrightarrow \cos A = -\frac{26}{70} = -0{\cdot}3714 \quad \therefore \quad A = 180° - 68°12' = 111°48'$

Substitute the numbers first, make cos A the subject of the formula. The ratio is negative, so angle A is obtuse. Subtract the angle found in the tables from 180°. A calculator gives the angle 111°18′ directly.

Chapter 16
Calculus

Differentiation

Figure 84(a) shows the graph of $f(x) = x^2$ or $y = x^2$ for positive values of x. To find the **gradient** of the curve at a point $P(x, y)$, x is given a small increase h as shown and the gradient of the chord PQ is found. (The diagram is enlarged for clarity.)

$QN = f(x+h)$ and $PM = f(x)$ $\therefore$ $QR = f(x+h) - f(x)$

gradient of chord $PQ = \dfrac{f(x+h)-f(x)}{h} = \dfrac{(x+h)^2-x^2}{h}$

$\dfrac{x^2+2xh+h^2-x^2}{h} \Leftrightarrow \dfrac{2xh+h^2}{h} \Leftrightarrow 2x + h$

If h is made smaller, Q moves towards P and the chord tends towards the position of the tangent at P, i.e. if $h \to 0$ the gradient of the chord $\to$ the gradient of the tangent. ($\to$ means 'tends to'.)

If $h \to 0$ then $2x+h \to 2x$. $2x$ is the gradient of the tangent at (x, y).

The symbol for the **gradient of the tangent is** $\dfrac{dy}{dx}$

$\therefore$ if $y = x^2$, $\dfrac{dy}{dx} = 2x$ also if $y = x^3$, $\dfrac{dy}{dx} = 3x^2$

and if $y = x^n$, $\dfrac{dy}{dx} = nx^{n-1}$

Note that (i) $y = 5x$, $\dfrac{dy}{dx} = 5$; (ii) $y = 4$, $\dfrac{dy}{dx} = 0$.

(i) $y = 5x$ is a straight line with constant gradient 5; (ii) is a horizontal line with zero gradient.

If $y = 4x^2$, $\dfrac{dy}{dx} = 4 \times 2x = 8x$. The curve $y = 4x^2$ has gradient 4 times the gradient of $y = x^2$.

This process of finding the gradient is called **differentiation**.

If $y = \dfrac{1}{x^2}$ write it as $y = x^{-2}$ then

$\dfrac{dy}{dx} = -2x^{-2-1} = -2x^{-3} = \dfrac{-2}{x^3}$

If $y = x^3 + 4x^2 + 5x + 2$, $\dfrac{dy}{dx} = 3x^2 + 8x + 5$

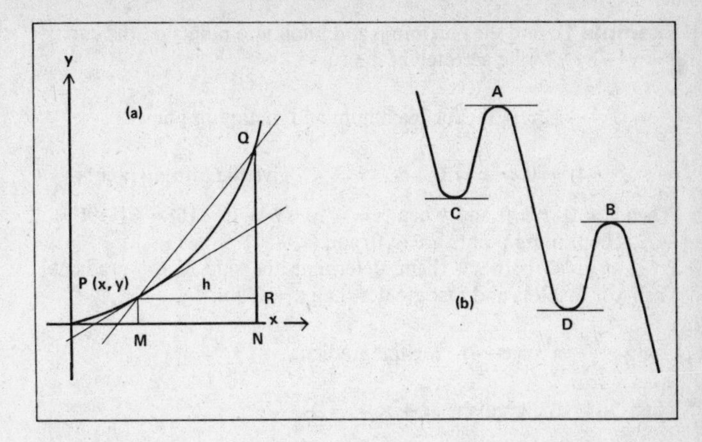

Figure 84

The differential of a sum = the sum of the differentials.

All of these results can be proved in a similar way to that of $y = x^2$. They give the exact gradient at any point, unlike the approximate method of drawing the graph shown on page 61.

Example Find the gradients at the points where $x = 2$ on

(i) $y = 3x + \frac{2}{x}$; (ii) $y = x^2 + 5x + 4$.

(i) write as $y = 3x + 2x^{-1}$

$$\frac{dy}{dx} = 3 - 2x^{-2}$$

at $x = 2$ the gradient = $3 - 2(2)^{-2} = 3 - 2 \times \frac{1}{4} = 2{\cdot}5$

(ii) $y = x^2 + 5x + 4$

$$\frac{dy}{dx} = 2x + 5$$

$\therefore$ at $x = 2$ the gradient = $4 + 5 = 9$.

Maximum and minimum

In figure 84(b) the points A and B are called the **maximum** points, and C and D are called the **minimum** points of the curve. There are greater and smaller values of y than these, but these are the names given to the turning points on the curve. It can be seen that the gradient is 0 at these points, i.e. $\frac{dy}{dx} = 0$. This fact helps us to find these points on a curve.

Example To find the maximum and minimum points on the curve $y = x^3 - 6x^2$. Make a sketch of the curve.

$$\frac{dy}{dx} = 3x^2 - 12x = 0 \quad \text{for maximum and minimum points.}$$

$$\Leftrightarrow 3x(x-4) = 0 \Leftrightarrow x = 0 \quad \text{or} \quad x = 4 \quad \text{give the turning points.}$$

When $x = 0$, $y = 0$ and when $x = 4$, $y = 64-(6\times16) = 64-96 = -32$. The turning points are $(0, 0)$ and $(4, -32)$.
To distinguish between them, determine the sign of the gradient when x is just less and just greater than $x = 0$ and 4.

Using $\dfrac{dy}{dx} = 3x(x-4)$ for the gradient,

x just > 4 the gradient $= + \times + = +$

x just < 4 the gradient $= + \times - = -$

The gradient like this forms a minimum point.

x just > 0 the gradient $= + \times - = -$

x just < 0 the gradient $= - \times - = +$

The gradient like this forms a maximum point.
$\therefore$ $(0, 0)$ is a maximum point and $(4, -32)$ is a minimum point.

To sketch the curve (i) plot these points; (ii) where possible find the points where the curve cuts the axes.
When $x = 0$, $y = 0$; this point has already been found.

When $y = 0$, $x^3 - 6x^2 = 0 \Leftrightarrow x^2(x-6) = 0 \Leftrightarrow x = 0$ or 6.

$\therefore$ the curve cuts the axes at $(0, 0)$ and $(6, 0)$.
There are no other turning points so the curve can be sketched as in figure 85. This and the curve on page 62 are typical cubic function curves.

The graph on page 62 is of $y = \frac{1}{2}(6+7x-x^3)$. We can check the accuracy of the maximum and minimum points.

$$\frac{dy}{dx} = \tfrac{1}{2}(7-3x^2) = 0 \quad \text{for maximum and minimum points.}$$

$$\Leftrightarrow 7 - 3x^2 = 0 \Leftrightarrow 3x^2 = 7 \Leftrightarrow x^2 = \tfrac{7}{3} = 2{\cdot}333 \Leftrightarrow x = \pm 1{\cdot}53$$

These are the same results as found by drawing, but remember that we can take a more accurate reading of $\sqrt{2{\cdot}3}$ if we wish.

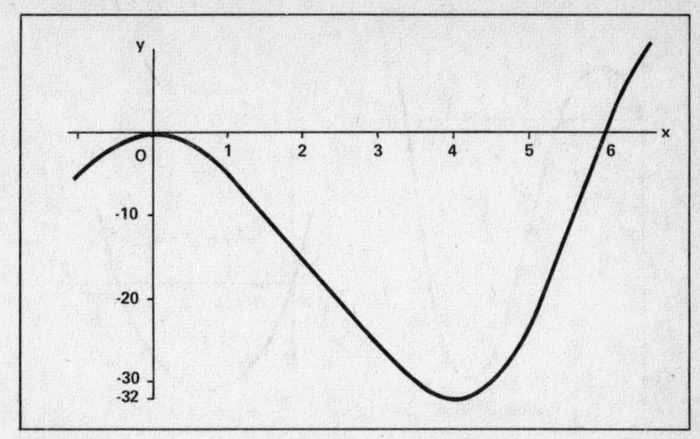

Figure 85

Rates of change

On page 58 it was established that the **gradient** at a point on a curve $y = f(x)$ measures the **rate** of **change** of y with respect to x. If the distance s from a fixed point is given in terms of t, i.e. $s = f(t)$ then the gradient on the curve is written $\frac{ds}{dt}$. This measures the rate of change of distance with respect to time, the speed or velocity.

If the velocity v is given in terms of t, i.e. $v = f(t)$, then the gradient is written $\frac{dv}{dt}$. This measures the rate of change of velocity, the acceleration.

$\boldsymbol{\frac{ds}{dt}}$ **gives velocity and** $\boldsymbol{\frac{dv}{dt}}$ **gives acceleration**

Example A body moves so that its distance in metres from a point 0 after t seconds is $s = t^3 - 9t^2 + 15t$. Find (i) the velocity when $t = 0$ and $t = 2$ seconds; (ii) when the velocity is zero; (iii) the acceleration when $t = 0$ and $t = 4$ seconds.

(i) the velocity $\frac{ds}{dt} = 3t^2 - 18t + 15$ differentiating.
when $t = 0$, $v = +15$m/s; when $t = 2$, $v = 12 - 36 + 15 = -9$m/s
a **negative** velocity signifies the **reverse** direction.

(ii) $v = 0$ when $3t^2 - 18t + 15 = 0 \Leftrightarrow 3(t^2 - 6t + 5) = 0$
$\Leftrightarrow 3(t-5)(t-1) = 0 \Leftrightarrow t = 5$ or $t = 1$
the velocity is zero after 1 s and after 5 s.

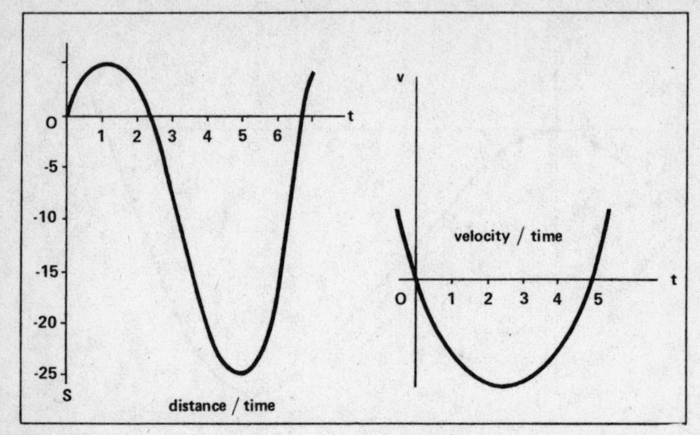

Figure 86

(iii) the acceleration $\frac{dv}{dt} = 6t - 18$ differentiating again; when $t = 0$ $a = -18\text{m/s}^2$ and when $t = 4$ $a = 24 - 18 = +6\text{m/s}^2$. The **negative** acceleration signifies **retardation**. See figure 86 for the distance/time, velocity/time graphs.

Integration

If $\frac{dy}{dx} = 3x^2$ then $y = \frac{3x^{2+1}}{2+1} = \frac{3x^3}{3} = x^3$. We use a process opposite to differentiation, i.e. add 1 to the index and divide by the new index. But $y = x^3 + 4$ and $y = x^3 - 56$ both have $\frac{dy}{dx} = 3x^2$.

The constant disappears on differentiating, so if $\frac{dy}{dx} = 3x^2$ we write $y = x^3 + c$. This is called the integral of $3x^2$, the process is called **integration** and c is the integration constant.

In general if $\boldsymbol{\frac{dy}{dx} = x^n}$ then $\boldsymbol{y = \frac{x^{n+1}}{n+1} + c}$

E.g. If $\frac{dy}{dx} = 4x \quad y = \frac{4x^{1+1}}{1+1} = \frac{4x^2}{2} \quad \therefore \quad y = 2x^2 + c$

If $\frac{dy}{dx} = 3$ then $y = 3x + c$, a straight line.

If $\frac{dy}{dx} = \frac{1}{x^2}$ write it as x^{-2} then $y = \frac{x^{-2+1}}{-2+1} = \frac{x^{-1}}{-1}$

$$\therefore \quad y = -\frac{1}{x} + c$$

If $\dfrac{dy}{dx} = 3x^2 + 4x + 3$ then $y = x^3 + 2x^2 + 3x + c$

The integral of a sum is the sum of the separate integrals.

To find the constant c, a point on the curve $y = f(x)$ is required.

Example The gradient at any point on a curve is $8x + 5$. The curve passes through the point $(1, 3)$. Find its equation.

$$\frac{dy}{dx} = 8x + 5 \quad \therefore \quad y = \frac{8x^{1+1}}{1+1} + 5x + c \Leftrightarrow y = 4x^2 + 5x + c$$

It passes through $(1, 3)$ $\therefore$ $3 = 4 + 5 + c \Leftrightarrow c = -6$
The required equation is $y = 4x^2 + 5x - 6$.

Integration can be used in velocity and acceleration. E.g. if a body moves so that its velocity after t seconds is $v = 2t + 3$ this means that $\dfrac{ds}{dt} = 2t + 3$. By integrating the distance $s = t^2 + 3t + c$ is found. If the body starts from O when $t = 0$, then $s = 0$ when $t = 0$. $\therefore$ $0 = c$ and $s = t^2 + 3t$ is the distance formula.

Integrating velocity gives distance and integrating acceleration gives velocity.

The area under the curve

To find the **area** cut off by the **curve** $y = x^2$, the x axis and the lines $x = 1$ and $x = 3$, as shown in figure 87(a), proceed as follows:

(1) integrate the function $\dfrac{x^3}{3} + c$

(2) substitute $x = 3$ and $x = 1$ $(\frac{27}{3} + c)$ and $(\frac{1}{3} + c)$

(3) subtract the results giving $\frac{26}{3}$ units^2, the required area.

The proof of this is omitted, but once again this gives the **exact** area and not an approximation as in the trapezium rule. The following example shows a better presentation.

Example Find the area cut off by the curve $y = 3x^2 - 6x$, the x axis and the lines $x = 4$ and $x = 2$ as shown in figure 87(b).

Integrate $[x^3 - 3x^2 + c]_2^4 = (64 - 48 + c) - (8 - 12 + c)$
$= (16 + c) - (-4 + c) = 20$. The area $= 20$ units^2.

The square brackets tell us that we are to substitute the two values and subtract. The larger is at the top and we always work out top minus bottom.

We are sometimes asked to find the area enclosed by the curve and the x axis. To find the limits, put $y = 0$:

$3x^2 - 6x = 0 \Leftrightarrow 3x(x-2) = 0 \Leftrightarrow x = 0$ or $x = 2$ are the limits.

Substitute these in the integral obtained above:

$[x^3 - 3x^2 + c]_0^2 = (8 - 12 + c) - (0 + c) = -4$ units^2.

The **negative** area denotes that it lies **below** the x axis.

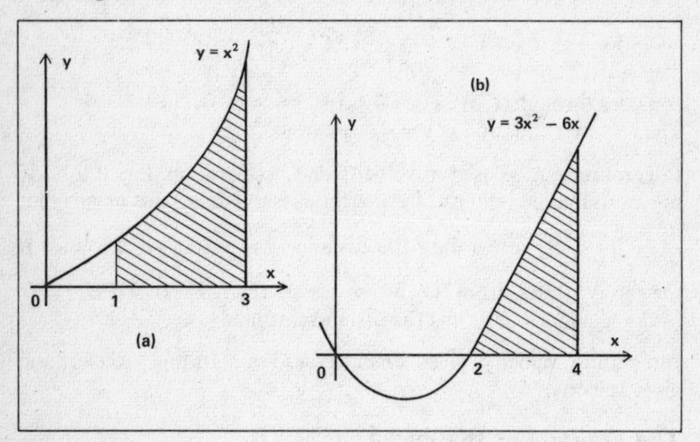

Figure 87

Key terms

Differentiation is the process of finding the gradient at any point on a curve. The result is called the differential coefficient.
The **maximum** and **minimum** points are the turning points or stationary points on the curve. At these points the gradient = 0.
Integration is the inverse process of differentiation. Given the gradient at any point, integration is used to find the equation of the curve. The result is called the indefinite integral.
We also use the process of integration to find the exact **area under a curve**, substituting the necessary limits. The result, a number, is called the **definite integral**.
Differentiation measures gradients, and is therefore a **rate measurer**. Velocities and accelerations are found by differentiating. **Integration** is also used to calculate velocities and distances.

Index

Examination Hints

This revision course contains examples similar to those you will meet in the examination. It is not enough just to read through the bookwork and its examples. After studying a topic and the examples provided, try to do them without looking at the book. If you cannot, return to the text and study it again. Repeat this until you succeed. Now pick out some questions on this topic from past examination papers. By working through these, you will gain valuable experience in the exact type of question you will meet in your own examination. Reserve one complete examination paper for a trial under examination conditions near the time of the examination itself.

In most GCE and CSE mathematics examinations now, the use of a calculator, logarithms or a slide rule is permitted in part or all of the examination. In most cases the tables are 4 figure, but those of the School Mathematical Project (SMP) are 3 figure, and contain a list of relevant formulae. Make sure that you know exactly where such aids are permitted and become thoroughly familiar with your calculator, logarithms or slide rule. There are some processes on the calculator that are not easy to carry out if you have not practiced them regularly. For example, to find the angle of which the cosine is $\frac{2}{7}$, in degrees and minutes. (The correct answer is 73°24′), or to substitute $x=-3$ in the expression $y=5+\dfrac{2}{x}-3x^2$. (The correct answer is $-22{\cdot}67$ to two decimal places.)

There is a growing practice in mathematics examinations of issuing formulae sheets to candidates, or printing the required formulae at the end of the relevant question. The formulae are provided only to confirm what you should already know. You cannot expect to use formulae effectively if you are not already familiar with them, so learn all the basic formulae thoroughly. Include in this the trigonometrical ratios: if you select the wrong ratio the rest of your answer will be of little use.

Show all your working in the examination. Even in multiple-choice questions you are not discouraged from doing so. Marks can be gained for method as well as for accuracy, so your incorrect answer can be worth more than half marks if you have shown a correct method. You are communicating your knowledge to somebody unknown to you who will judge you on what he sees in front of him. Make sure he understands clearly what you are trying to say.

Keep this in mind as you practise, and develop a neat style as you go. Write only one statement per line. Your working will be easier for you to check and for the examiner to follow. Keep rough and incidental working clear from the main arguments and statements. The right-hand side of the page is the best place for rough work. See the model answers on pages 189–90.

As a final note on revision, do not leave it until the last minute or you may find that it is too late to overcome your difficulties. Regular practice will help you to consolidate the basic ideas, which again will increase you confidence.

When you receive your examination paper, read the instructions carefully and adhere to them strictly. Read each question through to the end before you attempt to answer it. The last line sometimes contains a special instruction or a vital piece of information without which you will not be able to answer the question.

If you are about to make a calculation involving a calculator etc., make a quick check of your working up to that point to see that the figures you are about to use are correct. Make a rough estimate of the answer. It is possible to press the wrong button on a calculator and obtain an absurd answer.

Do not struggle over a question too long if you find it difficult: pass on to the next and return to it later if it is compulsory.

Finally, a word of advice to those students who have to take two equally-balanced papers. After the first paper, make a list of all the topics which have not been tested. They are almost certain to come up on the second paper.

You must know the metric system thoroughly. Mistakes are frequently made in changing units of area and volume. To change cm^2 to m^2, divide by 100^2, not 100. To change cm^3 to mm^3, multiply by 10^3, not 10. Do not treat your final calculations casually. A slide rule does not mean that you can take 3·84 as 4. The tables do not give you the complete answer. E.g. $\sqrt{0{\cdot}9} \neq 0{\cdot}3$ but $= 0{\cdot}9487$, and $(23{\cdot}1)^2 \neq 5336$ but $= 533{\cdot}6$. Make your rough estimate, then make a careful calculation or reference to the tables. Learn the basic processes in fractions. Fraction questions as such do not often appear in the papers, but without them you could find difficulties with questions on similar triangles, enlargements, matrices, trigonometry, statistics and others.

One of the aims of the new approach to mathematics is to make students more aware of the structure of the subject, so that elementary mistakes will be eliminated. Unfortunately in algebra the traditional errors are still being made.
$5x^2 \times 2x^3 \neq 7x^5$ or $10x^6$, but $= 10x^5$. (See page 18.)
$(a+b)^2 \neq a^2+b^2$. Multiply out the brackets $(a+b)$ $(a+b)$ to find the correct answer: $a^2+2ab+b^2$.
When factorising an expression such as $x^2-3x-10$, students tend to make two particular errors: (i) the signs in the brackets become confused: $(x+5)(x-2)$ is incorrect; $(x-5)(x+2)$ is correct; (ii) the working is pursued further to give $x = 5$ or $x = -2$. The solution of the equation $x^2-3x-10 = 0$ is confused with the factorising of $x^2-3x-10$. When confronted with a fraction such as $\frac{4+x}{4}$ do **not** cancel the 4's.
If $x^2 = 25$ remember that $x = -5$ as well as $+5$. $x^2 = 25 \Leftrightarrow x = \pm 5$. $-x > 5 \Leftrightarrow x < -5$. Remember to change the arrow with the sign.

In questions on sets, do not assume information to be exclusive. In the one on page 15, 18 boys like cricket. This does not mean that they like only cricket. The 18 includes those who like cricket and tennis, cricket and swimming and all three sports.

Can you produce the inverse of a 2×2 matrix with ease? Questions involving matrices are often well answered up until the point at which the inverse is required. Either learn the four steps listed on page 76, or be able to produce it from first principles.

Do know your trigonometry ratios? You cannot expect to obtain even method marks if you cannot produce the correct ratio at the right time.

O-level questions are usually arranged to work out conveniently if no errors are made. Just one small error could make a considerable difference to the time you spend on a question. Most of the mistakes are fairly trivial in themselves and with regular checking can be eliminated.

Model answers
In some questions you may be asked to answer 'true' or 'false' to a series of statements, or to pick out those which are correct.

Example Which of the following pairs of inequalities represent the shaded area, including the boundaries, in figure 88(a)?

(a) $y \geqslant x \quad y \leqslant 4$ (b) $x \leqslant 4 \quad x+y \leqslant 4$
(c) $y \geqslant x \quad x+y \geqslant 4$ (d) $y \leqslant x \quad x+y \geqslant 4$

Find the equations of the lines and hence the inequalities:

The line through (0, 0), (4, 4) is $y = x$ $\therefore$ $y \geqslant x$ in the shaded area.
The line through (0, 4), (4, 0) is $x+y = 4$ $\therefore$ $x+y \geqslant 4$ in the shaded area.
The line through (0, 4), (4, 4) is $y = 4$ $\therefore$ $y \leqslant 4$ in the shaded area.
The line through (4, 0), (4, 4) is $x = 4$ $\therefore$ $x \leqslant 4$ in the shaded area.

It can be seen that (a) and (c) are true, but (b) and (d) are false.

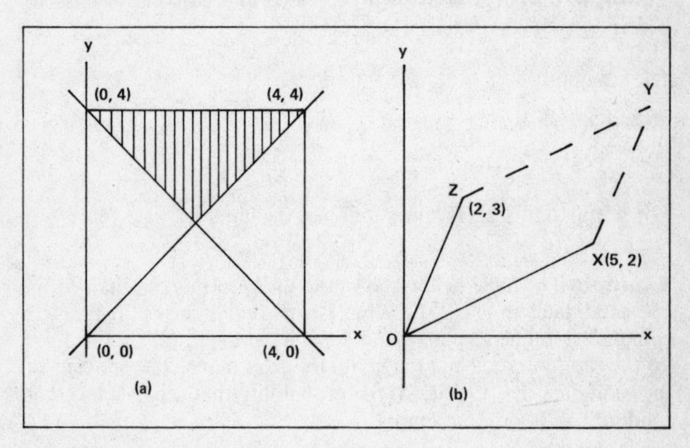

Figure 88

In a true multiple-choice paper each question is followed by 5 answers. Only one of them is correct. Do not guess the answer. Work out your answer in the usual way and compare it with those given.

Example $OXYZ$ is a parallelogram. O is (0, 0); X is (5, 2); Z is (2, 3). The co-ordinates of Y are (A) (3, –1); (B) $(3\frac{1}{2}, 2\frac{1}{2})$; (C) (10, 6); (D) (8, 5); (E) (7, 5).

Sketch a diagram as in figure 88(b), noting the order of the letters. The translation $O \to Z$ is the same as that of $X \to Y$.

$O \to Z$ is $\begin{pmatrix} 2 \\ 3 \end{pmatrix}$ $\therefore$ Y is the point $\begin{pmatrix} 5 \\ 2 \end{pmatrix} + \begin{pmatrix} 2 \\ 3 \end{pmatrix} = \begin{pmatrix} 7 \\ 5 \end{pmatrix}$

It can be seen that (E) is the correct answer.

Example An object travels a distance of 225 m in 25 seconds. Its speed in km/hour is (A) 9; (B) 32·4; (C) 41; (D) 28·2; (E) 110.

In 1 second it travels $\frac{225}{25}$ m; in 1 hour $\frac{225}{25} \times 60 \times 60$ m $= 9 \times 60 \times 60 = 32\,400$ m. Change to km. 32·4 km/hr is the speed, so (B) is the correct answer.

Example 0·15 in base 8 when expressed as a fraction in base 10 is: (A) $\frac{13}{512}$; (B) $\frac{15}{512}$; (C) $\frac{15}{100}$; (D) $\frac{13}{64}$; (E) $\frac{15}{64}$.

The column headings in base 8 are $\frac{1}{8}, \frac{1}{64} \ldots$

$0{\cdot}15 = \frac{1}{8} + \frac{5}{64} = \frac{8}{64} + \frac{5}{64} = \frac{13}{64}$.

(D) is the correct answer.

For setting out ordinary questions, see the hints on page 186.

Example The table below shows the grouped marks obtained by 150 candidates in a test. (i) What is the modal class? (ii) Draw a cumulative-frequency curve and from it estimate: (a) the median; (b) the semi-interquartile range; (c) the pass mark if 30 per cent of the candidates are to fail; (d) the probability that a paper taken at random has 38 or more marks.

mark	1–10	11–20	21–30	31–40	41–50
frequency	20	30	55	35	10

(i) The modal class is 21–30 marks.

(ii)

mark not more than:	10	20	30	40	50
cumulative frequency	20	50	105	140	150

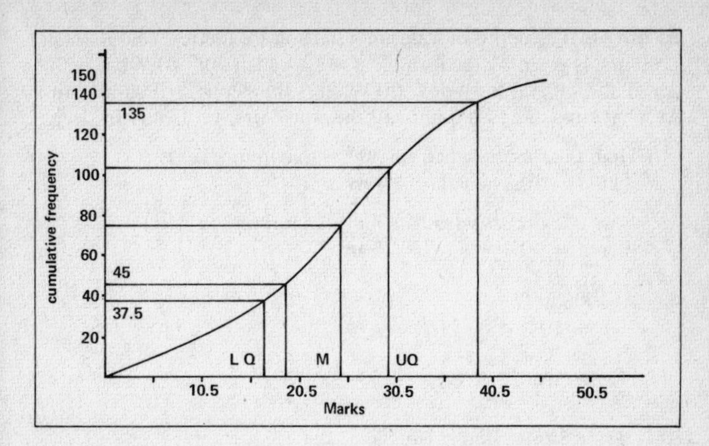

Figure 89

(a) The median mark = 25 (the $75\frac{1}{2}$th candidate)
(b) Upper quartile = 29·5 marks. Lower quartile = 16·75 marks
The semi-interquartile range $= \frac{1}{2}(29{\cdot}5 - 16{\cdot}75)$
$= 6{\cdot}38$

29·5
16·75
2 | 12·75
6·375

(c) $\frac{30}{100} \times 150 = \frac{4\,500}{100} = 45$ candidates

The 45th candidate has 18 marks $\therefore$ 19 is the pass mark.
(d) The 135th candidate scores 38 marks.
Candidates with 38 or more $= 150 - 134 = 16$.
the probability $= \frac{16}{150}$ or 0·11 (by slide rule)

Example Solve (i) the equation $x^2 - 3x - 54 = 0$; (ii) the inequation $11 - 2x > x - 4$.

(i) $x^2 - 3x - 54 = 0$
$\Leftrightarrow (x-9)(x+6) = 0$
$\Leftrightarrow x - 9 = 0 \quad \text{or} \quad x + 6 = 0$
$\Leftrightarrow$ **$x = 9$ or -6**

54 × 1 27 × 2
(9 × 6) 18 × 3

(ii) $11 - 2x > x - 4$
$\Leftrightarrow -2x - x > -4 - 11$
$\Leftrightarrow -3x > -15$
$\Leftrightarrow x < 5$

Example The depth of water in a harbour, d metres, t hours after 1200 hrs, is given by the formula $d = 10+4 \sin 30t°$. (i) What is the depth at high and low tides? (ii) What is the depth at 1900 hrs? (iii) At what time after 1200 hrs will the water first be 13·464 m deep?

(i) High tide occurs when $\sin 30t°$ is maximum, i.e. 1
then $d = 10 + 4 \times 1 =$ **14 m**

Low tide occurs when $\sin 30t°$ is minimum, i.e. -1
then $d = 10 + 4 \times -1 =$ **6 m**

(ii) At 1900 hrs $t = 7$

$$\therefore d = 10 + 4 \sin 210°$$
$$= 10 + 4 \times -\tfrac{1}{2}$$
$$= 10 - 2 = \mathbf{8\ m}$$

(iii) $\Leftrightarrow 13{\cdot}464 = 10 + 4 \sin 30t$

$\Leftrightarrow 3{\cdot}464 = 4 \sin 30t°$

$$\Leftrightarrow \sin 30t° = \frac{3{\cdot}464}{4}$$

$4\,|\,3{\cdot}464$ = 0·8660

$\Leftrightarrow \sin 30t° = 0{\cdot}8660$

$\Leftrightarrow 30t° = 60°$

$\Leftrightarrow t = 2$ the **time will be 1400 hrs**.

Examination papers

Booklets of past papers can be ordered from local bookshops or from the appropriate address below:

The School Mathematics Project (SMP)
Oxford and Cambridge Schools Examination Board, 10 Trumpington Street, Cambridge; and Elsefield Way, Oxford.

Mathematics in Education and Industry (MEI).
The Director, 57 High Street, Harrow, Middx., HA1 3HT.

Associated Examining Board (AEB)
The Secretary, Wellington House, Station Road, Aldershot, Hants., GU11 1BQ.

London (Syllabus C) Publications Office, 52 Gordon Square, London, WC1H 0P5.

MME and JMB (Syllabus C) J. Sherratt & Sons Ltd., Park Road, Timperley, Altrincham, Cheshire, WA14 5QQ.

Cambridge (Syllabus C) and Oxford (051 or 052) Educational Supply Association, The Pinnacles, Harlow, Essex, CM19 5AY.

Key Facts Revision Section

Chapter 1. Sets

A set is a well-defined collection of objects. E.g. the vowels $V = \{a, e, i, o, u\}$ or the natural numbers $N = \{1, 2, 3, 4, 5 \ldots\}$. Alternatively the set $A = \{5, 10, 15, 20\}$ can be written in the form $A = \{n : n \text{ a multiple of } 5, 1 \leqslant n \leqslant 20\}$.

Definitions

A **Venn diagram** is a pictorial representation of sets.
The **universal set** $\mathscr{E}$ is the set of reference. On a Venn diagram it is shown by a rectangle. See figure 1, page 9.
The symbol $\in$ denotes that an element belongs to a set, and $\notin$ denotes that it does not. E.g. if $A = \{1, 2, 3, 4\}$ then $4 \in A$, $7 \notin A$.
Equal sets have the same elements irrespective of the order in which they are listed. E.g. if $A = \{p, q, r\}$, $B = \{q, r, p\}$ then $A = B$.
$\subset$ denotes 'is a **subset** of'. If $B = \{1, 3, 5\}$ and $A = \{1, 2, 3, 5, 7\}$ then $B \subset A$. It can also be written $A \supset B$. See figure 1(a), page 9, for the Venn diagram showing one set inside another.
The **empty set** has no elements. Use the symbol $\{\ \}$ or $\varnothing$.
If $A = \{a, b, c\}$ and $D = \{b, c, d, e\}$ the **intersection** of these sets is the set containing the elements both in A **and** in D. $A \cap D = \{b, c\}$. In figure 1(b), page 9, it is the area shaded both ways.
The **union** of the same sets $A \cup D = \{a, b, c, d, e\}$, the set containing the elements in **either** A **or** D or **both**, each element listed once. In figure 1(b) it is the total area shaded in A or D.
The **complement** of a set A, written A', is the set formed by the elements of $\mathscr{E}$ not in A. E.g. If $\mathscr{E} = \{0, 2, 4, 6, 8\}$ and $A = \{2, 6\}$ then $A' = \{0, 4, 8\}$. See figure 1(c), page 9, for the Venn diagram.
Disjoint sets have no common elements. If $A = \{2, 6\}$ and $F = \{1, 3\}$ then $A \cap F = \varnothing$. The intersection is the empty set. See figure 1(d).
If $V = \{\text{vowels}\}$ then $\boldsymbol{n(V)} = 5$, i.e. the number of elements in set V.

If $\mathscr{E}$ is the universal set and A is a given set then $\mathscr{E} \cup A = \mathscr{E}$, $\mathscr{E} \cap A = A$, $A \cup A = A$, $A \cap A = A$, $A \cup A' = \mathscr{E}$, $A \cap A' = \varnothing$.
If B is a subset of A then $A \cup B = A$ and $A \cap B = B$.
The empty set is a subset of every set. $\varnothing \subset A$, $\varnothing \subset B \ldots$.

The laws of sets

The **commutative law** $A \cap B = B \cap A$: $A \cup B = B \cup A$.
The **associative law** $(A \cup B) \cup C = A \cup (B \cup C)$;
$(A \cap B) \cap C = A \cap (B \cap C)$
The **distributive law** (i) $A \cap (B \cup C) = (A \cap B) \cup (A \cap C)$
(ii) $A \cup (B \cap C) = (A \cup B) \cap (A \cup C)$

The Venn diagrams showing the associative law are in figures 1(e)(f) and (g)(h) respectively; the distributive law is shown (i) in figures 2(a)(b) and (ii) in figures 2(c)(d).

De Morgan's laws $(A \cap B)' = A' \cup B'$; $(A \cup B)' = A' \cap B'$. These are shown in figures 3(a) (b) and 3(c) (d) respectively.

In addition to the examples in chapter 1, there are others on pages 66 and 69.

Chapter 2. Basic algebra

Directed numbers

$(+3)+(+2)=+5$ $(+3)-(+2)=+1$ $+\times+=+$ $+\div+=+$
$(+3)+(-2)=+1$ $(+3)-(-2)=+5$ $+\times-=-$ $+\div-=-$
$(-3)+(+2)=-1$ $(-3)-(+2)=-5$ $-\times+=-$ $-\div+=-$
$(-3)+(-2)=-5$ $(-3)-(-2)=-1$ $-\times-=+$ $-\div-=+$

Note that $+(+2)=+2$: $+(-2)=-2$: $-(+2)=-2$: $-(-2)=+2$.

When **multiplying** or **dividing** two like signs the answer will be plus. For two unlike signs the answer will be minus.

The laws of indices

$x^3 = x \times x \times x$. The 3 is called the **power** of x. x is the **base**.

$a^m \times a^n = a^{m+n}$; $a^m \div a^n = a^{m-n}$: $(a^m)^n = a^{mn}$.

To multiply numbers with the same base add the indices; to divide them subtract the indices; and to raise them to a given power multiply the indices.

E.g. $2^3 \times 2^2 = 2^5 = 32$: $X^5 \div X^2 = X^3$: $(2^4)^3 = 2^{12} = 4096$.

The laws of algebra

(i) $a+b=b+a$: $ab=ba$; (ii) $(a+b)+c=a+(b+c)$: $(ab)c=a(bc)$; (iii) $a(b+c)=ab+ac$.
(i) the **commutative law**; (ii) the **associative law**; (iii) the **distributive law**.

Like terms

$3xy+4xy=7xy$, but $3x^2+2xy$ cannot be added further. We can only add and subtract **like** terms. x^2 is not 'like' xy. But $3x^2 \times 2xy = 6x^3y$: the terms do not have to be like for multiplication and division. Note that the indices on the x and y terms are not added; x and y are different bases.

Brackets

$a(x+y)=ax+ay$ and $x(x+3y+z)=x^2+3xy+xz$ are examples of the distributive law. To multiply a bracket by a single term multiply every term inside the bracket by the term outside.

$(a+b)(c+d) = ac+ad+bc+bd$. Multiply each term in the first bracket by each term in the second bracket.

E.g. $(x+3)(x+2) = x^2+2x+3x+6 = x^2+5x+6$.
The expression reduces to three terms. An expression of the form ax^2+bx+c is called a **quadratic** expression.

Factors

In algebra the factors of an expression multiply out to give the expression. Here are some examples:
$3x^2-18x = 3x(x-6)$. Note that $3x$ has been distributed.
$ax+2bx-3a-6b = x(a+2b)-3(a+2b) = (a+2b)(x-3)$. Factorise in pairs, to find a common bracket of $(a+2b)$.

E.g. $x^2-3x-10$ (i) Write down two brackets thus: $(x \quad)(x \quad)$; (ii) find two numbers with product -10 and sum -3; they are -5 and $+2$; $-5\times+2=-10$ and $-5+2=-3$. The factors are $(x-5)(x+2)$. Similarly $x^2-7x+12 = (x-4)(x-3)$ because $-4\times-3=+12$ and $-4-3=-7$.

The special cases:
$a^2+2ab+b^2 = (a+b)(a+b) = (a+b)^2$ a perfect square.
$a^2-2ab+b^2 = (a-b)(a-b) = (a-b)^2$ a perfect square.
$a^2-b^2 = (a-b)(a+b)$ the difference between two squares.
a^2+b^2, the sum of two squares, has no factors.

In the case of $4x^2-4x-15$, write down all the combinations of $(4x \quad 15)(x \quad 1)$; $(4x \quad 5)(x \quad 3)$; $(2x \quad 5)(2x \quad 3)$, and multiply them out to find that the factors are $(2x-5)(2x+3)$.

Fractions

To simplify:

$$\text{(i)} \quad \tfrac{1}{4}(4x+8y) = \frac{4(x+2y)}{4} = x+2y$$

$$\text{(ii)} \quad \frac{5}{2x} - \frac{4}{3x^2} = \frac{5\times 3x}{6x^2} - \frac{4\times 2}{6x^2} = \frac{15x-8}{6x^2}$$

(i) Factorise and cancel where possible. (ii) Put both fractions over the common denominator $6x^2$. We treat them as ordinary fractions in arithmetic.

Equations

Solve the **simple** equation $\tfrac{1}{4}(x-2)+\tfrac{1}{3}(x-3) = 2$.
$\Leftrightarrow 3(x-2)+4(x-3) = 24 \Leftrightarrow 3x-6+4x-12 = 24 \Leftrightarrow 7x-18 = 24$
$\Leftrightarrow 7x = 24+18 \Leftrightarrow 7x = 42 \Leftrightarrow x = 6$ is the **single** solution.
Multiply through by the common denominator 12, remove the brackets and rearrange the terms to find x.

To solve the **simultaneous** equations

$5x - 3y = 14$ (i) $\times 2 \Rightarrow 10x - 6y = 28$
$7x + 2y = 1$ (ii) $\times 3 \Rightarrow 21x + 6y = 3$
adding $31x = 31 \Rightarrow x = 1$

substitute in (ii) $7 + 2y = 1 \Rightarrow 2y = -6 \Rightarrow y = -3$
(i) $\times 2 +$ (ii) $\times 3$ eliminates the y terms. $(1, -3)$ is the solution. (i) and (ii) are straight lines. They intersect at $(1, -3)$.

To solve the **quadratic** equation $2x^2 - 5x = 7 \Leftrightarrow 2x^2 - 5x - 7 = 0 \Leftrightarrow (2x-7)(x+1) = 0 \Leftrightarrow 2x - 7 = 0$ and $x + 1 = 0 \Leftrightarrow 2x = 7$ or $x = -1 \Leftrightarrow x = 3\frac{1}{2}$ or -1, the two **solutions** or **roots** of the equation. (i) Make right-hand side $= 0$; (ii) factorise; (iii) $ab = 0 \Rightarrow a = 0$ or $b = 0$.

Two cases worth special note:

(i) $3x^2 = 75$
$\Leftrightarrow 3x^2 - 75 = 0$
$\Leftrightarrow 3(x^2 - 25) = 0$
$\Leftrightarrow 3(x-5)(x+5) = 0$
$\Leftrightarrow x = 5$ or -5

(ii) $3x^2 = 75x$
$\Leftrightarrow 3x^2 - 75x = 0$
$\Leftrightarrow 3x(x - 25) = 0$
$\Leftrightarrow x = 0$ or $x = 25$

(i) contains a difference between two squares; in (ii) each term has a common factor of $3x$.

To solve $2x^2 - 4x - 1 = 0$ to 2 decimal places. This does not factorise. Using the formula from page 26: $a = 2$ $b = -4$ $c = -1$

$$x = \frac{-(-4) \pm \sqrt{(-4)^2 - 4 \times 2 \times -1}}{4} = \frac{+4 \pm \sqrt{16+8}}{4} = \frac{4 \pm \sqrt{24}}{4}$$

$$x = \frac{4 + 4{\cdot}899}{4} \text{ or } \frac{4 - 4{\cdot}899}{4} \quad \therefore \quad x = \frac{8{\cdot}899}{4} \text{ or } \frac{-0{\cdot}899}{4}$$

$\therefore \quad x = 2{\cdot}22$ or $-0{\cdot}22$ to 2 decimal places.

Rearrangement of formulae

To make x the subject of the formula $A = \dfrac{x+y}{x-y}$

$\Leftrightarrow A(x-y) = (x+y) \Leftrightarrow Ax - Ay = x + y \Leftrightarrow Ax - x = Ay + y$

$\Leftrightarrow x(A-1) = Ay + y \Leftrightarrow x = \dfrac{Ay+y}{A-1}$. x is now the subject.

Using inverses (i) multiply both sides by $x - y$; (ii) add Ay to both sides and subtract x from both sides; (iii) divide both sides by $(A-1)$. In other examples, be prepared where necessary to square or square-root both sides. See page 26.

$\frac{a}{b} = \frac{c}{d} \Leftrightarrow ad = bc$. This is called **cross-multiplying**. It can be useful when there is only **one** fraction on each side of the = sign.

Chapter 3. Number systems

Integers = $\{\ldots -2, -1, 0, 1, 2, 3 \ldots\}$
Natural numbers = $\{1, 2, 3, 4 \ldots\}$
Whole numbers = $\{0, 1, 2, 3, 4 \ldots\}$
Rational numbers = $\{\frac{1}{3}, 0{\cdot}7, \frac{3}{4} \ldots\}$
Irrational numbers = $\{\pi, \sqrt{2}, \sqrt{3}, \sqrt{5} \ldots\}$
Prime numbers = $\{2, 3, 5, 7 \ldots\}$
All the above sets together form the set of **real numbers**. Rational numbers can be expressed as a fraction, or a decimal which terminates or recurs. Irrational numbers cannot be expressed in this form.

Different bases

Base	Column headings							14 in these bases
10 **Denary**			...	1000	100	10	1	14
2 **Binary**	...	32	16	8	4	2	1	1110
3	...	243	81	27	9	3	1	112
4		...	256	64	16	4	1	32
8 **Octal**			...	512	64	8	1	16

To convert (i) 228 (denary) to base 4; (ii) 134_5 to denary.

(i)

4	228
4	57 r 0
4	14 r 1
4	3 r 2
4	0 r 3

$228_{10} = 3210_4$

(ii)

25	5	1
1	3	4

$= (25 \times 1) + (5 \times 3) + (1 \times 4)$
$= 25 + 15 + 4 = 44$

$134_5 = 44_{10}$

(i) Carry out successive division by 4. The remainders give the answer. Note the order. (ii) Put down the base 5 headings, the 134 in the correct columns and find the total as shown.

Note The largest digit required is one less than the base in which you are working.

Calculations

(i) $122_3 + 221_3$

$$\begin{array}{r} 122 \\ +\ 221 \\ \hline 1120 \\ \hline \end{array}$$

(ii) $1110_2 - 111_2$

$$\begin{array}{r} 1\ 1\ 1\ 0 \\ -\ \ 1\ 1\ 1 \\ \hline 1\ 1\ 1 \\ \hline \end{array}$$

Notice the carry and borrow process in each base (i) in base 3, $2+1=10$ put down 0 carry 1; (ii) in binary $0-1$ will not go, borrow 2 from the next column.

(iii) Work out $52_8 \times 13_8$

$$\begin{array}{r} 52 \\ \times\ 13 \\ \hline 176 \\ 52 \\ \hline 716 \end{array}$$

(iv) Evaluate $x^2 - x$ if $x = 11_5$; answer in base 5.

$x^2 - x = x(x-1)$

$= 11(11-1)$

$= 11 \times 10 = 110$

Answer $= 110_5$.

Modulo arithmetic

Modulo 3 assumes that only $\{0, 1, 2\}$ exists, modulo 5 has only $\{0, 1, 2, 3, 4\}$. Calculate on a clock face as on page 35. Use your pen as a pointer. In modulo 5, $2+3=0$. Start with 2 units, **add** 3 units and point at 0. $2 \times 3 = 1$. Take 2 units three **times**, to point at 1. $1-2=4$, point at 1, **go back** 2 units to point at 4. The complete addition and multiplication tables in modulo 5 are:

+	0	1	2	3	4
0	0	1	2	3	4
1	1	2	3	4	0
2	2	3	4	0	1
3	3	4	0	1	2
4	4	0	1	2	3

×	0	1	2	3	4
0	0	0	0	0	0
1	0	1	2	3	4
2	0	2	4	1	3
3	0	3	1	4	2
4	0	4	3	2	1

Binary operations

A **binary operation** is carried out between two members of a set S. The symbol used is $*$, where $*$ can be $+$, $-$, $\times$, $\div$, $()^2$ etc. For the following definitions let $S = \{0, 1, 2, 3, 4\}$ and let $*$ be addition, i.e. addition in modulo 5. See the table above for the results.

Closure Each element of the table $\in S$. We say that set S is closed under the operation $*$.

Identity elements Note that $0+0=0$; $0+1=1+0=1$; $0+2=2+0=2$, etc. The value of an element is unaltered when 0 is added. We say that S possesses an identity element under $*$. It is 0. Notice that $0 \in S$. This is important. See 'group', below.

Inverse elements Notice that $1+4=4+1=0$: $2+3=3+2=0$ and $0+0=0$. When 4 is added to 1 the answer is 0, the identity element. 4 is called the inverse element of 1 under $*$. Every element of S possesses an inverse element under $*$. Notice that each inverse is also a member of S. This is important. See 'group', below.

Associativity If $(a*b)*c = a*(b*c)$ for all elements of S under $*$, S is associative under $*$. Consider $(2+3)+4 = 0+4 = 4$ and $2+(3+4) = 2+2 = 4$. It is true for these three elements and is also true for every other three elements of S.

Group If a set S under an operation $*$ is (i) closed; (ii) has an identity element which is a member of S; (iii) each term has an inverse which is also in S; (iv) is associative, the set S forms a group under $*$. Notice that modulo 5 forms a group under addition. Notice that modulo 5 does not form a group under multiplication. The inverses of the elements do not belong to S. Isomorphic groups are groups with the same structure, they 'work' in the same way. See page 38.

Chapter 4. Functions (1)

A **relation** connects members of two sets. Figure 8(a), page 41, shows the relation 'is a prime factor of', between $\{2, 3, 5\}$ called the **domain** and $\{4, 5, 6, 8\}$ called the **range**. This is a **mapping**, each member of the domain mapping on to one or more of the range. If only one arrow leaves each member of the domain, the relation is called a **function**, see figure 8(b), and if one arrow goes to each member of the range, the function has **1 to 1 correspondence**, see figure 8(c).

Algebraic functions

If 4 is added to x then x becomes $x+4$. Write this $x \to x+4$. For each value of x there is only one value $x+4$, so this relation is a function. Write it $f : x \to x+4$ or $f(x) = x+4$.

E.g. $f : x \to 2x-1$ The domain is $\{0, 2, 4\}$. Find the range.
We have $f(x) = 2x-1$ when $x = 0$ $f(0) = 0-1 = -1$
when $x = 2$ $f(2) = 4-1 = 3$ when $x = 4$ $f(4) = 8-1 = 7$.
The range is $\{-1, 3, 7\}$. This can be shown on a mapping diagram similar to the one on page 42. Notice that $f(2)$ means substitute $x = 2$ in the given function.

Compound functions

$f : x \to 3x-2$ and $g : x \to x^2+1$ are two functions. $fg(x)$ or fg is the compound function 'g followed by f'. If $x = 4$ $fg(4)$ is found as follows: (i) find $g(4) = 16+1 = 17$; (ii) now find $f(17) = (3 \times 17) - 2 = 49$. $\therefore fg(4) = 49$.
Compare this with $gf(4)$: (i) find $f(4) = 12-2 = 10$; (ii) now find $g(10) = 100+1 = 101$. $\therefore gf(4) = 101$. $fg \neq gf$. Observe the 'followed by' carefully. If we have $fgh(x)$ then find 'h followed by g followed by f'.

Inverse functions

If $f : x \to x+5$ the inverse function $f^{-1} : x \to x-5$ maps the range on to the domain. See figure 10(b), page 44. Some other inverses are

$$f : x \to x-2 \quad f^{-1} : x \to x+2, \quad f : x \to 3x \quad f^{-1} : x \to \frac{x}{3}$$

E.g. If $f: x \to \sqrt{x}$ find $f^{-1}(4)$. $f^{-1}: x \to x^2$ is the inverse function. $f^{-1}(4) = (4)^2 = 16$.

For less obvious inverses we can use a **flow diagram**. If $f: x \to 4x-3$, to find $f^{-1}(5)$ we write out the process of finding f for a given x as a flow chart:

Take x → [×4] → $4x$ → [−3] → $f(x) = 4x - 3$

for f^{-1} use inverses in the reverse order

$f^{-1} = \dfrac{x+3}{4}$ ← [÷4] ← $x + 3$ ← [+3] ← Take x

$f^{-1}(x) = \frac{1}{4}(x+3) \quad \therefore \quad f^{-1}(5) = \frac{1}{4}(5+3) = 2.$

Note that (i) the inverse of $\dfrac{1}{x}$ is $\dfrac{1}{x}$; (ii) $ff^{-1}(x) = x$ for all x.

Flow charts can also help in rearranging formulae, as on page 46.

Co-ordinates

A point in a plane is positioned by its **co-ordinates** (x, y) relative to the x and y axes. See figure 11(a), page 48.

To find the distance PQ between $P(1,2)$ and $Q(5,3)$. In figure 11 form right-angled $\triangle PQR$.

QR = difference between the y's: PR = difference between the x's. $QR = 3-2 = 1$; $PR = 5-1 = 4$.

By Pythagoras' theorem $PQ = \sqrt{(4^2+1^2)} = \sqrt{17} = 4{\cdot}12$.

Gradient of $PQ = \dfrac{\text{vertical}}{\text{horizontal}} = \dfrac{QR}{PR} = \dfrac{\text{difference of } y\text{'s}}{\text{difference of } x\text{'s}} = \dfrac{1}{4}$

This is **tan *QPR***. A negative gradient gives a backwards slope.

The straight line

$y = mx + c$ is the equation of a **straight line** with gradient m passing through the y axis at the point $(0, c)$. In figure 12, page 49, (i) $y = 3$; (ii) $x = 2$; (iii) $y = x$; (iv) $y = -x$ are the lines: (i) horizontal through $(0, 3)$; (ii) vertical through $(2, 0)$; (iii) at 45° with both axes; (iv) at 45° backwards with both axes. See figure 12(b) and 13(a), page 49.

To interpret the line $6x - 2y = 7$. Rearrange in the form above $-2y = -6x + 7 \Leftrightarrow y = 3x - 3{\cdot}5 \therefore m = 3$ and it cuts the y axis at the point $(0, -3{\cdot}5)$.

To find the equation of the line with slope -3 passing through $(2, 1)$. $m = -3$, the line is $y = -3x + c$, passing through $(2, 1) \therefore 1 = -3 \times 2 + c \Leftrightarrow c = 7$. The line is $y = -3x + 7$.

To find where the line $3x+5y=15$ cuts the axes (the **intercepts**); put $x=0$; $5y=15 \Leftrightarrow y=3$. So the line cuts the y axis at $(0,3)$. Put $y=0$; $3x=15 \Leftrightarrow x=5$. So the line cuts the x axis at $(5,0)$.

To find the equation of the line joining the two points $A(-3,5)$ and $B(2,1)$ we require figure 14(b), page 51. Follow this and the working on page 51.

These are some of the ways of using $y=mx+c$. Use the one which seems most convenient at the time.

Chapter 5. Functions (2)

Figure 16 on page 53 shows the graphs of $y=x^2$, $y=x^2-2$, $y=-x^2$, $y=2-x^2$. The y axis is the axis of symmetry in each case. The curve is called a **parabola**, as is any curve of the form $y=ax^2+bx+c$.

To sketch the curve of $y=x^2-3x+2$. Find its intersections with the axes. When $x=0$, $y=2$. When $y=0$ we have $x^2-3x+2=0$. $\Leftrightarrow (x-2)(x-1)=0 \Leftrightarrow x=2$ or 1. We now have three points $(0,2)$ $(2,0)$, $(1,0)$ and can sketch the parabola as in figure 17(a). The axis of symmetry is the line $x=1{\cdot}5$.

Figure 17 also shows the curves of $y=x^3$ and $y=\frac{1}{x}$. Knowing these shapes will be useful for the following section, on variation.

Variation

If $y=\{2,4,6,8\}$ (range) when $x=\{1,2,3,4\}$ (domain), y is twice x. y varies directly as x, or y **varies as** x, in symbols: $\boldsymbol{y \propto x}$ or $\boldsymbol{y=kx}$. The graph is a straight line through $(0,0)$ with slope k.
If $y=\{1,4,9,16\}$ when $x=\{1,2,3,4\}$ then y **varies** as the **square** of x. $\boldsymbol{y \propto x^2}$ or $\boldsymbol{y=kx^2}$. The graph is a parabola similar to figure 16(a).
If y **varies** as the **cube** of x we write $\boldsymbol{y \propto x^3}$ or $\boldsymbol{y=kx^3}$. The graph is similar to figure 17(c).

E.g. y varies as the square of x. $x=2$ when $y=36$. Find y when $x=5$. We know that $y=kx^2$ and we find k by substituting the known values. $36=4k \Leftrightarrow k=9$. $y=9x^2$ is the curve required. When $x=5$, $y=(9\times 25)=225$, $y=225$ is the required answer.

If $y=\{1,\frac{1}{2},\frac{1}{3},\frac{1}{4}\}$ when $x=\{1,2,3,4\}$, y varies as the **inverse** of x.

$\boldsymbol{y \propto \frac{1}{x}}$ or $\boldsymbol{y=\frac{k}{x}}$. The graph is similar to figure 17(d). $y=\frac{k}{x^2}$ denotes that y varies as the inverse of the square of x.

Joint variation occurs when y depends on two other variables. E.g. If $y = 3xz^2$, y varies jointly as x and the square of z.

E.g. p varies jointly as q and the inverse of r. $r = 4$, $p = 3$ when $q = 2$. Find r when $q = 5$ and $p = 2$.

We know that $p = \frac{kq}{r}$ $\therefore$ $3 = \frac{k \times 2}{4} \Leftrightarrow 2k = 12 \Leftrightarrow k = 6$

$\therefore$ $p = \frac{6q}{r}$ is the curve. $2 = \frac{6 \times 5}{r} \Leftrightarrow 2r = 30 \Leftrightarrow r = 15$.

In (i) $C = 2\pi r$; (ii) $A = \pi r^2$; (iii) $V = \pi r^2 h$, we have (i) C varies as r; (ii) A varies as r^2; (iii) V varies jointly as h and the square of r.

Rates of change

Figure 18(b) on page 58 shows a distance/time graph. The **slope of OA** $= \frac{\text{vertical}}{\text{horizontal}} = \frac{\textbf{distance}}{\textbf{time}}$ **= speed**. Speed is the rate of change of distance. The **gradient** measures the **rate of change**. The horizontal line AB signifies a zero velocity. In general on $y = mx + c$ the gradient m measures the rate of change of y with respect to x. It is constant. On a curve such as $y = x^2 + 1$ in figure 19(a), the slope varies at each point. It is measured by plotting the graph accurately (page 61) or by calculus (page 176).

Figure 19(b) shows a velocity/time graph (straight line). The gradient $= \frac{\textbf{velocity}}{\textbf{time}}$ **= acceleration**. The horizontal line AB denotes a zero acceleration, i.e. constant velocity. Figure 20 shows a velocity/time curve. The acceleration varies, maximum at $t = 2$, zero when $t = 4$.
Note that a negative velocity means the reverse direction, a negative acceleration means retardation. The following table summarises some of the information gained from these graphs.

graph	**type**	**velocity**	**acceleration**
distance/time	straight line	constant	zero
distance/time	curve	varies	constant, or varies
velocity/time	straight line	varies	constant
velocity/time	curve	varies	varies

The **area** under the velocity/time graph = **distance** travelled. To find the area under figure 19(b), divide it into known shapes, $\triangle$'s, etc., and find the exact area. In figure 20 draw vertical lines at equal intervals, and assume that the segments of curve cut off are straight lines. We have a series of **trapeziums**. Calculate the area of each to find the approximate distance travelled. The calculation is shown on page 60.

Plotting graphs accurately

(i) Make a clear table of values; (ii) use the scale given, where applicable; (iii) place the axes so that all the points can be plotted; (iv) plot the points and join them with a smooth curve; (v) label the axes and the curve. A straight line needs three points only. From the graph we read maximum and minimum points, solve equations, inequations, and measure slopes as shown on page 61.

Chapter 6. Inequalities

The signs $>$ greater than, $<$ less than, $\geqslant$ greater or equal to, $\leqslant$ less or equal to are used.
$4 > 2$ is a true statement. We can (i) add the same number to both sides, say 3, then $7 > 5$; (ii) subtract the same number from both sides, say 4, then $0 > -2$; (iii) multiply and divide both sides by the same + number, say $+3$, then $12 > 6$. But when we multiply or divide by a minus number we must reverse the arrow to make the statement true. E.g. Multiplying by -2, $-8 < -4$ is true. Apply these to the solution of inequations.

E.g. $3(x+3) > 4x+5 \Leftrightarrow 3x+9 > 4x+5 \Leftrightarrow 3x-4x > 5-9$ $\Leftrightarrow -x > -4 \Leftrightarrow x < 4$ is the solution set.

To solve $x^2-5x+4 < 0$ it is better to sketch the graph of $y = x^2-5x+4$ and read the solution from there.
When $x = 0$ $y = 4$ when $y = 0$ $x^2-5x+4 = 0 \Leftrightarrow (x-4)(x-1) = 0$ $\Leftrightarrow x = 4$ or 1. Plot the parabola through the points $(0, 4)(4, 0)(1, 0)$. The area below the x axis provides the solution $4 > x > 1$.

Locus

A **locus** is the path or area traced out by a point moving to a certain rule. E.g. The locus of a point P moving so that its distance from two fixed pobets A and B is such that $PA < PB$ is the area shaded in figure 23(c). The broken line denotes that P does not lie on the boundary.

These ideas are used to define areas on the co-ordinate plane. Consider the shaded areas in figure 24, page 67.
(a) is $x \geqslant 0$, $y \geqslant 0$, $x < 3$, $y \leqslant 2$. Notice the broken line.
(b) is $y \leqslant x$, $y \geqslant -x$. Note, this automatically makes $x \geqslant 0$.
(c) is the line $x+y = 5$. For points above the line $x+y > 5$, but below the line $x+y < 5$, which is the same as $y < 5-x$.
In set notation we have $P = \{(x, y) : x \geqslant 0, y \geqslant 0\}$ and $P = \{(x, y) : x+y \leqslant 5\}$ which says that P must lie on or inside the triangle with vertices $(0, 0)$, $(5, 0)$, $(0, 5)$. See figure 24(c).

Linear programming

This is the process of reducing a problem to terms of inequalities, plotting them and making the required deductions from the graphs. It is important to be able to write down the necessary inequalities.

E.g. A shop keeps two types of article, A and B, in stock. It always keeps at least 10 of A and 16 of B. A costs £10 and B £15. The insurance covers no more than £600 value and the shop always stocks more than 30 articles. Let x be the number of A and y be the number of B in stock, so the four inequalities are: (i) $x \geqslant 10$ (ii) $y \geqslant 16$ (iii) $x+y > 30$ (iv) $10x+15y \leqslant 600$.

The four lines $x = 10$, $y = 16$, $x+y = 30$ and $10x+15y = 600$ can be plotted. The area enclosed by them will satisfy all the conditions. x and y are whole numbers, and it is the whole-number values of (x, y) which are of interest here. For a complete example, see page 70.

Chapter 7. Matrices

The **order** of a matrix is the number of rows × number of columns.

If $A = \begin{pmatrix} 3 & 4 \\ 1 & 2 \end{pmatrix}$ $B = \begin{pmatrix} 5 & 7 \\ 6 & 0 \end{pmatrix}$ $C = \begin{pmatrix} 2 \\ 1 \end{pmatrix}$

multiply AB

$$\begin{pmatrix} 3 & 4 \\ 1 & 2 \end{pmatrix} \times \begin{pmatrix} 5 & 7 \\ 6 & 0 \end{pmatrix} = \begin{pmatrix} 3\times5+4\times6 & 3\times7+4\times0 \\ 5\times1+2\times6 & 1\times7+2\times0 \end{pmatrix}$$

$$= \begin{pmatrix} 39 & 21 \\ 17 & 7 \end{pmatrix}$$

$$BA = \begin{pmatrix} 22 & 34 \\ 18 & 24 \end{pmatrix}$$

$$BC = \begin{pmatrix} 5 & 7 \\ 6 & 0 \end{pmatrix}\begin{pmatrix} 2 \\ 1 \end{pmatrix} = \begin{pmatrix} 10+7 \\ 12+0 \end{pmatrix} = \begin{pmatrix} 17 \\ 12 \end{pmatrix}$$

$CB = \begin{pmatrix} 2 \\ 1 \end{pmatrix}\begin{pmatrix} 5 & 7 \\ 6 & 0 \end{pmatrix}$ which cannot be evaluated.

Note (i) In general $AB \neq BA$ (ii) CB and CA do not evaluate. The number of columns of the left matrix must equal the number of rows of the right matrix. (iii) The order of the answer is:

(number of rows of the left) × (number of columns of the right).

(iv) In AB, B is pre-multiplied by A or A is post-multiplied by B.

Adding $A + B = \begin{pmatrix} 3+5 & 4+7 \\ 1+6 & 2+0 \end{pmatrix} = \begin{pmatrix} 8 & 11 \\ 7 & 2 \end{pmatrix}$ A and B must have the same order for + and −.

$k\begin{pmatrix} a & b \\ c & d \end{pmatrix} = \begin{pmatrix} ka & kb \\ kc & kd \end{pmatrix}$ also note $\begin{pmatrix} 9 & 15 \\ 6 & 12 \end{pmatrix} = 3\begin{pmatrix} 3 & 5 \\ 2 & 4 \end{pmatrix}$

unit matrix $I = \begin{pmatrix} 1 & 0 \\ 0 & 1 \end{pmatrix}$ or $\begin{pmatrix} 1 & 0 & 0 \\ 0 & 1 & 0 \\ 0 & 0 & 1 \end{pmatrix}$

$$IA = \begin{pmatrix} 1 & 0 \\ 0 & 1 \end{pmatrix}\begin{pmatrix} 3 & 4 \\ 1 & 2 \end{pmatrix} = \begin{pmatrix} 3 & 4 \\ 1 & 2 \end{pmatrix} = A$$

also $AI = A$. I must be square, i.e. a matrix of order 2×2, 3×3 etc.

Zero or **null** matrix $\begin{pmatrix} 2 & -1 \\ 0 & 0 \end{pmatrix} \times \begin{pmatrix} 3 & 0 \\ 6 & 0 \end{pmatrix} = \begin{pmatrix} 0 & 0 \\ 0 & 0 \end{pmatrix} = O$ the zero matrix, all elements zero.

Note $AI = IA = A$ and $A + O = O + A = A$. I and O are **identity** matrices.

Transpose of A is $A' = \begin{pmatrix} 3 & 1 \\ 4 & 2 \end{pmatrix}$. The rows and columns are changed.

The **determinant** of $\begin{pmatrix} a & b \\ c & d \end{pmatrix}$ is the value of $ad - bc$.

The **inverse** A^{-1} of A is such that $A^{-1} \times A = A \times A^{-1} = I$, the unit matrix. To find the inverse of A (a 2×2 matrix) proceed as follows: (i) find the determinant of A; $3 \times 2 - 4 \times 1 = 2$; (ii) swap the top left and bottom right elements; (iii) change the sign of top right and bottom left elements; (iv) divide each element by the determinant.

(ii) (iii) $\begin{pmatrix} 2 & -4 \\ -1 & 3 \end{pmatrix}$

(iv) $\begin{pmatrix} \frac{2}{2} & -\frac{4}{2} \\ -\frac{1}{2} & \frac{3}{2} \end{pmatrix} = \frac{1}{2}\begin{pmatrix} 2 & -4 \\ -1 & 3 \end{pmatrix} = A^{-1}$

The inverse is used to solve simultaneous equations. For the proof of the above process and its use see page 76. The multiplication and addition of other orders is shown on pages 72, 73 and 74.

Chapter 8. Elementary geometry

Angles

A is **acute** if $0° < A < 90°$; **obtuse** if $90° < A < 180°$; **reflex** if $180° < A < 360°$. A and B are **complementary** if $A + B = 90°$, **supplementary** if $A + B = 180°$. In figure 28(d), page 81, l_1 and l_2 are parallel. $a = b$ (alternate), $a = c$ (corresponding), $a + d = 180°$ (interior), $b = c$ (vertically opposite).

Triangles

The sum of the angles = 180°. An exterior angle = the sum of the interior opposite pair, see figure 29(a).

A **scalene** triangle has no two sides or angles equal. An **isosceles** triangle has two sides and two angles equal. An **equilateral** triangle has all sides and angles equal (60°).

Congruent figures have the same shape and size; **similar** figures have the same shape, but differ in size. In **similar** triangles the corresponding sides are in the same ratio, and the angles of each are equal. A calculation is shown on page 83. If the ratio of the sides = $\mathbf{1 : n}$, then the ratio of the areas = $\mathbf{1 : n^2}$ and for similar solids the ratio of volumes = $\mathbf{1 : n^3}$.

In figure 31(a) **Pythagoras'** theorem states that in right-angled triangle ABC $a^2 = b^2 + c^2$. (i) $c^2 = a^2 - b^2$ or (ii) $b^2 = a^2 - c^2$.

Use (i) and (ii) when the hypotenuse is known, to find another side. Remember the 3 : 4 : 5, 5 : 12 : 13, 8 : 15 : 17, 7 : 24 : 25 right-angled triangles and the 6 : 8 : 10, $2\frac{1}{2} : 6 : 6\frac{1}{2}$ ones etc.

Polygons

A **quadrilateral** has 4 sides, a **pentagon** 5 sides, a **hexagon** 6 sides, an **octagon** 8 sides.

The sum of the exterior angles of an n-sided polygon = 360°.

The sum of the interior angles = $(2n-4) \times 90°$.

A **regular** polygon has all sides and angles equal. The interior angle of a regular hexagon = 120°, the exterior = 60°. For a regular pentagon the interior = 108°, the exterior = 72°. See figure 32.

Quadrilaterals

Sum of interior angles = 360°. A **trapezium** has two parallel sides. A **kite** has two pairs of adjacent sides equal, and its diagonals at 90°. A **parallelogram** has opposite sides equal and parallel; the diagonals bisect each other. A **rhombus** is a parallelogram with all sides equal; the diagonals bisect at 90°. A **rectangle** is a **parallelogram** with right angles; the diagonals are equal. A **square** is a rhombus with right angles. See figures 33 and 34.

Symmetry

If a figure **reflects** on to itself in a line it has **bilateral symmetry**. The line is a **line of symmetry**. If it **rotates** on to itself about an axis, it has **rotational symmetry**. E.g. The kite $ABCD$ in figure 33(b), page 88, has line symmetry in AC. The square in figure 34(d) has 4 lines of symmetry and can **rotate** on to its own shape **4 times in one revolution (360°)**. It has **order 4** rotational symmetry. The following table lists the symmetry properties of the basic figures. See figures 29, 32, 33, 34.

figure	lines of symmetry	rotational order
isosceles triangle	1	1 (once in 360°)
equilateral triangle	3	3 (every 120°)
kite	1	1 (once in 360°)
parallelogram	0	2 (every 180°)
rhombus	2	2 (every 180°)
rectangle	2	2 (every 180°)
square	4	4 (every 90°)
regular pentagon	5	5 (every 72°)
regular hexagon	6	6 (every 60°)

Circles

A **chord** cuts the circumference of a circle into the **major** and **minor arcs**, and the area into the **major** and **minor segments**. Theorems: (i) the perpendicular bisector of a chord passes through the centre of the circle; (ii) a radius and the tangent through the point of contact are at right angles; (iii) two tangents can be drawn to a circle from a point outside: they are equal in length; (iv) if two circles touch (a) externally (b) internally, the distance between their centres is (a) the sum (b) the difference of the radii. See figure 36.

The **angle** theorems: (i) the angle standing on an arc or chord at the centre of a circle = twice the angle standing on the arc or chord at the circumference; (ii) angles standing on the same arc are equal; (iii) the angle in a semi-circle = 90°; (iv) a cyclic quadrilateral has its 4 corners on a circle; the opposite angles of such a quadrilateral add up to 180°. See figure 37.

Chapter 9. Transformation geometry

In any transformation a point A **maps** on to its **image** A', $A \rightarrow A'$. **Rotation**, **reflection** and **translation** are **isometries**. The figure and its image are equal in size and shape (congruent). **Invariant** points and lines do not move in a transformation.

Reflection occurs in a line, the axis of reflection, which is invariant. It is an isometry. See figure 40.

In the co-ordinate plane if we reflect in the:
(i) x axis $(x, y) \rightarrow (x, -y)$; (ii) y axis $(x, y) \rightarrow (-x, y)$;
(iii) line $y = x (x, y) \rightarrow (y, x)$; (iv) line $y = -x (x, y) \rightarrow (-y, -x)$.

Translation All points move equal distances along parallel lines. No points are invariant. It is an isometry. See figure 39(b).

A translation of $\begin{pmatrix} 6 \\ 2 \end{pmatrix}$ maps (x, y) on to $(x+6, y+2)$, $+6$ in the direction of the x axis, followed by $+2$ in the y axis direction.

Rotation occurs about an axis, which is invariant, + direction anti-clockwise. It is an isometry.
We construct the centre and angle of rotation as shown on page 97.
In the co-ordinate plane a rotation about $(0, 0)$ of:
(i) $+90°$ $(x, y) \to (-y, x)$; (ii) $180°$ (either way) $(x, y) \to (-x, -y)$;
(iii) $+270°$ $(x, y) \to (y, -x)$; (iv) $-90°$ $(x, y) \to (y, -x)$.
Note that (iii) and (iv) are the same. See figure 42(a).

Enlargement requires a scale factor and a centre (which is invariant). The figure and image are similar. In figure 43 $\triangle A_1 B_1 C_1$ is an enlargement scale factor $+2$ in O of ABC.

$$\frac{OA_1}{OA} = \frac{OB_1}{OB} = \frac{OC_1}{OC} = \frac{2}{1}. \qquad \textbf{Area } \triangle A_1 B_1 C_1 = 4 \triangle ABC$$

When the enlargement is on the opposite side of O to $\triangle ABC$ the scale factor is negative. See $\triangle A_2 B_2 C_2$.

Shear The figure is pushed over. In figure 44(a) the base AB is invariant, the height and area are unaltered; only the shape changes.

Stretch Enlarges the figure in one direction only. See figure 44(b).

The product of transformations
If M = reflection, R = rotation, the product MR is the transformation 'R followed by M'. Do not assume that $MR = RM$. Keep to the order stated. If a figure returns to its original position it is called the **identity transformation *I***. E.g. If H = half turn then H^2 or $HH = I$.
In figure 45, H = half turn, M and N are reflections in m and n respectively. NMH is (1) H then (2) M then (3) N as follows

(1)	(2)	(3)	
D	D	B	so $NMH = I$
A C	C A	C A	
B	B	D	

Many products can be reduced to a single transformation. E.g. A reflection in the x axis (M) followed by a rotation of $+90°$ about $O(R)$ is identical to a reflection in the line $y = x(P)$ or $RM = P$. But R followed by M = reflection in $y = -x(Q)$ or $MR = Q$ $\therefore$ $MR \neq RM$.

Chapter 10. Transformation matrices

The point (x, y) when written $\begin{pmatrix} x \\ y \end{pmatrix}$ can be transformed by pre-multiplying it by one of the following (2×2) matrices.

Reflection

(i) $\begin{pmatrix} 1 & 0 \\ 0 & -1 \end{pmatrix}$ (ii) $\begin{pmatrix} -1 & 0 \\ 0 & 1 \end{pmatrix}$ (iii) $\begin{pmatrix} 0 & 1 \\ 1 & 0 \end{pmatrix}$ (iv) $\begin{pmatrix} 0 & -1 \\ -1 & 0 \end{pmatrix}$

The above matrices cause a reflection in (i) the x axis; (ii) the y axis; (iii) the line $y = x$; (iv) the line $y = -x$.

Rotation

(i) $\begin{pmatrix} 0 & -1 \\ 1 & 0 \end{pmatrix}$ (ii) $\begin{pmatrix} -1 & 0 \\ 0 & -1 \end{pmatrix}$ (iii) $\begin{pmatrix} 0 & 1 \\ -1 & 0 \end{pmatrix}$

The above matrices cause a point to rotate about $(0, 0)$ through (i) $+90°$ (ii) $180°$ either way (iii) $+270°$ (or $-90°$).

Enlargement

$\begin{pmatrix} k & 0 \\ 0 & k \end{pmatrix}$ causes an enlargement of scale factor k in $(0, 0)$.

Shear and stretch

(i) $\begin{pmatrix} 1 & a \\ 0 & 1 \end{pmatrix}$ (ii) $\begin{pmatrix} 1 & 0 \\ b & 1 \end{pmatrix}$ (iii) $\begin{pmatrix} a & 0 \\ 0 & 1 \end{pmatrix}$ (iv) $\begin{pmatrix} 1 & 0 \\ 0 & b \end{pmatrix}$

(i) causes a **shear** parallel to the x axis; (ii) a shear parallel to the y axis; (iii) causes a **stretch** parallel to the x axis; (iv) a stretch parallel to the y axis. a and b determine the length of the stretch or shear.

E.g. (1) Reflect (2, 1) in the y axis; (ii) rotate (2, 1) $+90°$ about (0, 0).

(i) $\begin{pmatrix} -1 & 0 \\ 0 & 1 \end{pmatrix}\begin{pmatrix} 2 \\ 1 \end{pmatrix} = \begin{pmatrix} -2 \\ 1 \end{pmatrix}$ $(2, 1) \rightarrow (-2, 1)$

(ii) $\begin{pmatrix} 0 & -1 \\ 1 & 0 \end{pmatrix}\begin{pmatrix} 2 \\ 1 \end{pmatrix} = \begin{pmatrix} -1 \\ 2 \end{pmatrix}$ $(2, 1) \rightarrow (-1, 2)$

Product of transformations

M = reflection in the x axis; R = rotation about $(0, 0)$ $+90°$. The product RM is found by multiplying the appropriate matrices.

$RM = \begin{pmatrix} 0 & -1 \\ 1 & 0 \end{pmatrix}\begin{pmatrix} 1 & 0 \\ 0 & -1 \end{pmatrix} = \begin{pmatrix} 0 & 1 \\ 1 & 0 \end{pmatrix}$ and

$$MR = \begin{pmatrix} 1 & 0 \\ 0 & -1 \end{pmatrix}\begin{pmatrix} 0 & -1 \\ 1 & 0 \end{pmatrix} = \begin{pmatrix} 0 & -1 \\ -1 & 0 \end{pmatrix}$$

RM = reflection in $y = x$. MR = reflection in $y = -x$. So $MR \neq RM$ as we saw under 'Product of Transformations', page 108.

E.g. If E = enlargement scale factor 2 in (0,0) find ME (2,3)

$$= \begin{pmatrix} 1 & 0 \\ 0 & -1 \end{pmatrix}\begin{pmatrix} 2 & 0 \\ 0 & 2 \end{pmatrix}\begin{pmatrix} 2 \\ 3 \end{pmatrix} = \begin{pmatrix} 2 & 0 \\ 0 & -2 \end{pmatrix}\begin{pmatrix} 2 \\ 3 \end{pmatrix} = \begin{pmatrix} 4 \\ -6 \end{pmatrix}:$$

$(2,3) \rightarrow (4,-6)$ Here $ME = EM$.

Note if H = half turn $H^2 = \begin{pmatrix} -1 & 0 \\ 0 & -1 \end{pmatrix}\begin{pmatrix} -1 & 0 \\ 0 & -1 \end{pmatrix} = \begin{pmatrix} 1 & 0 \\ 0 & 1 \end{pmatrix} = I$, the unit matrix or the **identity** transformation. A figure returns to its original position with a half turn followed by a half turn.

Chapter 11. Vectors

In figure 52(a) the **vector** XY, written $\overrightarrow{XY}$ or **a**, has magnitude and direction. The length XY represents the magnitude. Equal vectors have the same length, are parallel **and** in the same direction. E.g. If $XY = PQ$ and XY is parallel to PQ, with the arrows pointing the same way, then $\overrightarrow{XY} = \overrightarrow{PQ}$ or $\mathbf{a} = \mathbf{b}$. The vector $\overrightarrow{YX}$ has the same magnitude as $\overrightarrow{XY}$ and is parallel to $\overrightarrow{XY}$ but acts in the opposite direction: $\overrightarrow{YX} = -\overrightarrow{XY}$ or $\overrightarrow{YX} = -\mathbf{a}$.

The **modulus** of $\overrightarrow{XY}$, written $|\overrightarrow{XY}|$, is the length of XY, ignoring the direction. $|\overrightarrow{XY}| = |\overrightarrow{YX}|$ or $|\mathbf{a}| = |-\mathbf{a}|$.

Addition

In figure 52(b) $\overrightarrow{OQ} = \overrightarrow{OP} + \overrightarrow{PQ} = \mathbf{a} + \mathbf{b}$. They are in the same line.
In figure 53(a) $\overrightarrow{OQ} = \overrightarrow{OP} + \overrightarrow{PQ}$ or $\mathbf{c} = \mathbf{a} + \mathbf{b}$.
In figure 53(b) $\mathbf{x} = \mathbf{a} + \mathbf{b} + \mathbf{c}$, the sum of three vectors.
In figure 53(c) $\mathbf{a} + \mathbf{b} + \mathbf{c} = \mathbf{0}$ returning to O. This is called a **zero** vector.

Notice that the sum of vectors is also a vector.
In modulus form $|\overrightarrow{OQ}| = |\mathbf{a} + \mathbf{b}|$; $|\mathbf{c}| = |\mathbf{a} + \mathbf{b}|$; $|\mathbf{x}| = |\mathbf{a} + \mathbf{b} + \mathbf{c}|$; $|\mathbf{a} + \mathbf{b} + \mathbf{c}| = 0$.

The parallelogram law

In figure 54(b) the diagonal $\overrightarrow{OC} = \mathbf{a} + \mathbf{b} = \mathbf{b} + \mathbf{a}$.
In figure 54(c) reverse $\overrightarrow{OB}$ then $\overrightarrow{BO} = -\mathbf{b}$ and in $\triangle BOA$ diagonal $\overrightarrow{BA} = -\mathbf{b} + \mathbf{a} = \mathbf{a} - \mathbf{b}$ $\therefore$ $\overrightarrow{AB} = \mathbf{b} - \mathbf{a}$.

This is vector **subtraction**, obtained by **reversing** a vector. Note (i) vector addition is commutative $\mathbf{a}+\mathbf{b} = \mathbf{b}+\mathbf{a}$.
(ii) $|\overrightarrow{BA}| = |\overrightarrow{AB}|$ or $|\mathbf{a}-\mathbf{b}| = |\mathbf{b}-\mathbf{a}|$

Figure 55(c) shows that $(\mathbf{a}+\mathbf{b})+\mathbf{c} = \mathbf{a}+(\mathbf{b}+\mathbf{c})$; vector addition is associative.

Scalar multiplication
In figure 55(a) $\overrightarrow{OP} = \overrightarrow{PQ} = \overrightarrow{QR} = \mathbf{a}$. $\overrightarrow{OR} = \mathbf{a}+\mathbf{a}+\mathbf{a} = 3\mathbf{a}$.
$3\mathbf{a}$ has three times the magnitude of $\mathbf{a}$. $|3\mathbf{a}| = 3|\mathbf{a}|$.

Figure 55(b) shows that scalar multiplication is distributive:
$2\mathbf{a}+2\mathbf{b} = 2(\mathbf{a}+\mathbf{b})$.

Position vector
In figure 56(a) O is the origin. $\overrightarrow{OA}$ or $\mathbf{a}$ is the **position vector** of A referred to O. If the co-ordinates of A are (4, 3) then $\overrightarrow{OA} = \begin{pmatrix} 4 \\ 3 \end{pmatrix}$

To find $|\overrightarrow{OA}|$ use Pythagoras' theorem: $|\overrightarrow{OA}| = \sqrt{3^2+4^2} = 5$ units.

We can add or subtract position vectors. If A and B have position vectors $\mathbf{a} = \begin{pmatrix} 4 \\ 3 \end{pmatrix}$ $\mathbf{b} = \begin{pmatrix} 1 \\ 5 \end{pmatrix}$ then $\mathbf{a}+\mathbf{b} = \begin{pmatrix} 4 \\ 3 \end{pmatrix}+\begin{pmatrix} 1 \\ 5 \end{pmatrix} = \begin{pmatrix} 5 \\ 8 \end{pmatrix}$ and $\mathbf{a}-\mathbf{b} = \begin{pmatrix} 4 \\ 3 \end{pmatrix}-\begin{pmatrix} 1 \\ 5 \end{pmatrix} = \begin{pmatrix} 3 \\ -2 \end{pmatrix}$. Note that $\begin{pmatrix} 5 \\ 8 \end{pmatrix}$ is the position vector of the fourth point, C, of the parallelogram $OACB$. See figure 56(a).

Components
If $\mathbf{c} = \mathbf{a}+\mathbf{b}$, $\mathbf{a}$ and $\mathbf{b}$ are called the **components** of $\mathbf{c}$. It is convenient to take them at right angles as follows: Let $\mathbf{i}$ be a unit (i.e. modulus 1) vector along the x axis and $\mathbf{j}$ be a unit vector along the y axis. If P has position vector $\begin{pmatrix} 3 \\ 2 \end{pmatrix}$ then vector $\overrightarrow{OP}$ can be written $3\mathbf{i}+2\mathbf{j}$. See figure 56(b) in which $|\overrightarrow{OP}| = \sqrt{3^2+2^2} = \sqrt{13} = 3{\cdot}61$.

E.g. If $\mathbf{a} = 3\mathbf{i}-2\mathbf{j}$ and $\mathbf{b} = 4\mathbf{i}+3\mathbf{j}$ then the vector $2\mathbf{a}+3\mathbf{b} = 2(3\mathbf{i}-2\mathbf{j})+3(4\mathbf{i}+3\mathbf{j}) = 6\mathbf{i}-4\mathbf{j}+12\mathbf{i}+9\mathbf{j} = 18\mathbf{i}+5\mathbf{j}$.

Triangle of velocities
The bearing of the direction towards which a pilot points an aircraft is the **course**. The speed in still air is the **air speed**. The **wind** blows **from** a direction with a given speed. The resulting direction taken by the plane is the **track**, and its speed, the **ground speed**. Figure 59(b) shows the triangle of velocities for a flight.

Using 1 arrow for course, 2 arrows for wind, 3 arrows for track, then 1 arrow + 2 arrow = 3 arrow (vector addition).
For a boat sailing on a river or at sea, the 'course' denotes the same thing; for 'air speed' read 'speed in still water'; for 'wind' read 'current'; track and ground speed are the same.

Use calculation where the triangle has a right angle, otherwise use a scale drawing (unless told to calculate). See the examples on pages 122–5.

Chapter 12. Basic arithmetic

Factors 2, 3, 4, 6 are factors of 12. They all divide into 12 exactly. The **product** of 3 and 4 is 12, the result of multiplication. The **quotient** of $12 \div 6$ is 2, the result of division.

Fractions

A **fraction** is $\dfrac{\text{numerator}}{\text{denominator}}$, $\dfrac{N}{D}$. $\dfrac{12}{15} = \dfrac{3}{4}$ in lowest terms **cancelling** common factors in N and D. $\frac{8}{3}$ is an improper fraction $(N > D)$; $\frac{8}{3} = 2\frac{2}{3}$ and similarly $\frac{4}{1} = 4$.

E.g. (i) Work out $\frac{3}{4}+\frac{4}{5}+\frac{7}{10}$. Which is the largest of the three?

The common $D = 20$

$$\therefore \quad \frac{3\times 5}{4\times 5}+\frac{4\times 4}{5\times 4}+\frac{7\times 2}{10\times 2} = \frac{15}{20}+\frac{16}{20}+\frac{14}{20} = \frac{45}{20} = 2\tfrac{1}{4}$$

By writing them with the common D we can see that $\frac{16}{20}$ or $\frac{4}{5}$ is the largest.

(ii) Work out $3\frac{1}{5}\times 1\frac{3}{4}$. $\quad = \frac{16}{5}\times\frac{7}{4} = \frac{4}{5}\times\frac{7}{1} = \frac{28}{5} = 5\frac{3}{5}$.

(iii) Work out $2\frac{1}{5}\div\frac{3}{5}$. $\quad = \frac{11}{5}\div\frac{3}{5} = \frac{11}{5}\times\frac{5}{3} = \frac{11}{3} = 3\frac{2}{3}$.

Decimals

$0{\cdot}3 = \frac{3}{10}$; $\;0{\cdot}76 = \frac{76}{100}$; $\;0{\cdot}005 = \frac{5}{1000}$

To change $\frac{3}{7}$ to a decimal to 2 places: $\quad 7\,\underline{|\;3{\cdot}000\,0}$
$\qquad\qquad 0{\cdot}428$

Go to one extra place and round off: $\frac{3}{7} = 0{\cdot}43$.

$4{\cdot}62 \times 1\,000 = 4\,620$; $469 \div 1\,000 = 0{\cdot}469$. To multiply or divide by 1 000 move the decimal point 3 places right (to multiply) or left (to divide). Move the point one place for each 0.

$5{\cdot}427 = 5{\cdot}43$ to 2 **decimal places** $= 5{\cdot}4$ to 1 decimal place.
$5{\cdot}427 = 5{\cdot}43$ to 3 **significant figures** $= 5{\cdot}4$ to 2 significant figures.
In both add 1 if last digit > 5 Leave if < 5. Make it even if the last digit $= 5$.

No measurement is exact. 7·2 cm is given to the nearest mm. It can lie between 7·15 and 7·25. We have $7{\cdot}2 \pm 0{\cdot}05$. These are the **limits of accuracy**. Also 8 g lies between 7·5 and 8·5 i.e. $8 \pm 0{\cdot}5$ g.

Indices

Continuing from page 194

$$a^0 = 1 \quad a^{-1} = \frac{1}{a} \quad a^{-2} = \frac{1}{a^2} \quad a^{\frac{1}{2}} = \sqrt{a}$$

E.g. $17^0 = 1 \quad 4^{-1} = \frac{1}{4} \quad 3^{-2} = \frac{1}{9} \quad 81^{\frac{1}{2}} = 9 \quad 27^{\frac{2}{3}} = 9$

Standard form

or scientific notation expresses a number in the form $A \times 10^n$ where $1 < A < 10$ and n an integer. E.g. for large numbers $7\,360\,000\,000 = 7{\cdot}36 \times 10^9$; $3\frac{1}{2}$ million $= 3{\cdot}5 \times 10^6$. For small numbers $0{\cdot}000\,000\,53 = 5{\cdot}3 \times 10^{-7}$.

Use of tables

Make a reasoned estimate before looking at the tables. You have to put your own decimal point in most of them.

E.g. (i) Find $25{\cdot}74^2$.
Estimate $20^2 = 400$, $30^2 = 900$
Tables give 6625
Answer = 662·5

(ii) $0{\cdot}732^2$.
$0{\cdot}7^2 = 0{\cdot}49 \quad 0{\cdot}8^2 = 0{\cdot}64$
Tables give 5538
Answer = 0·5538

E.g. (i) Find $\sqrt{7620}$

Pair off from decimal point

|7 6|2 0|·

Nearest root below 76 is 8 ($8^2 = 64$)
First figure is 8; use 8729:

space out as

8	7 ·	2 9
76	20 ·	

Answer = 87·29

(ii) find $\sqrt{0{\cdot}0762}$

0·|0 7|6 2|

Nearest root below 7 is 2; use 2760:

·	2	7	6 0
0·	07	62	

Answer = 0·2760

E.g. Evaluate $\frac{1}{0{\cdot}046}$ Estimate $\approx \frac{1}{0{\cdot}05} = \frac{100}{5} = 20$.

From the reciprocal tables we find 2174. Answer = 21·74.

To carry out the longer calculations use a slide rule, or logs. A logarithm is an index. To multiply numbers add the logs; to divide numbers subtract the logs, obeying the laws of indices. A slide rule has log scales printed on it. To multiply numbers using a slide rule add log distances; to divide numbers subtract log distances, etc. Calculations using both are shown on pages 134–7. Do not forget to estimate first.

Metric system

Remember that kilo = 1 000, milli = $\frac{1}{1\,000}$ and centi = $\frac{1}{100}$. E.g. 1 km = 1 000 m; 1 kg = 1 000 grams (g); 1 m = 100 cm; 1 m = 1 000 mm.

When changing units in area use the square of the conversion. When changing units in volume use the cube of the conversion. E.g. 500 000 cm^2 = 50 m^2 ($\div 100 \times 100$). 2 $\text{km}^3 = 2 \times 10^9\ \text{m}^3$.

Learn the basic formulae, pages 138–9. Do not evaluate π until necessary. E.g. How many cylindrical cups of radius 3 cm, height 5 cm, can be filled from a cylindrical jug of radius 10 cm, height 18 cm?

Volume of cup $= \pi \times 3^2 \times 5$ Volume of jug $= \pi \times 10^2 \times 18$

$$\text{number of cups} = \frac{\pi \times 100 \times 18}{\pi \times 9 \times 5} = 40.$$

π has cancelled, no substitution is required.

Sector

$$\frac{\text{arc length}}{\text{circumference}} = \frac{\text{area of sector}}{\text{area of circle}} = \frac{\text{angle at centre}}{360^\circ}$$

See figure 63, page 141. These two formulae can be used to find the length of arc and area of a sector for any given angle in any circle with given radius.

Ratio

A ratio compares numbers. Two heights of 160 cm and 180 cm are in the ratio 160 : 180 = 8 : 9 in lowest terms. A ratio is a fraction (sin, cos, tan).

E.g. Divide a 48 cm rod into the ratio 3 : 4 : 5. It is divided into $3+4+5 = 12$ parts. Each part is $48 \div 12 = 4$ cm. $3 \times 4 = 12$ cm $4 \times 4 = 16$ cm, $5 \times 4 = 20$ cm. The pieces are 12, 16, 20 cm.

A **map** with scale 1 : 25 000 means that 1 cm on the map = 25 000 cm on the ground. 5 cm on the map = $5 \times 25\,000$ = 125 000 cm = 1 250 m = 1·25 km on the ground.
5 km on the ground = 500 000 cm. On the map $500\,000 \div 25\,000$ = 20 cm.

Percentage

$20\% = 0{\cdot}2$ or $\frac{1}{5}$; $7\% = 0{\cdot}07$ or $\frac{7}{100}$; $33\frac{1}{3}\% = 0{\cdot}\dot{3}$ or $\frac{1}{3}$; $250\% = 2{\cdot}5$ or $2\frac{1}{2}$.

E.g. To express 27 kg as a percentage of 72 kg, write as a fraction and multiply by 100. $\frac{27}{72} \times 100 = \frac{3}{8} \times 100 = \frac{300}{8} = 37{\cdot}5\%$.

E.g. Find 13% of 12: $\frac{13}{100} \times 12 = \frac{156}{100} = 1{\cdot}56$

(i) $\frac{100 + r}{100}$ (ii) $\frac{100 - r}{100}$

(i) will increase a number by $r\%$
(ii) will decrease a number by $r\%$

E.g. Increase £28 by 12%; $\frac{112}{100} \times 28 = \frac{3136}{100} = £31{\cdot}36$.

E.g. A number when increased by 12% becomes 28. Find the number.
Let x = the original number. The new number $\frac{112}{100}x = 28$

$\Leftrightarrow x = 28 \times \frac{100}{112}$ $\therefore$ x the original $= \frac{100}{4} = 25$.

Notice that the **original** is always **100%**.

Interest

Principal (P); Interest (I); Rate (R); Time (T); Amount (A).

simple interest = $\frac{PRT}{100}$. The interest each year is calculated on the original principal. In **compound interest** the interest is calculated on the amount $(P+I)$ at the end of each year. See the examples on pages 144–5.

Chapter 13. Statistics

Representing data

A **pie chart** shows the share-out of a given quantity. The 360° are divided into the same ratio as the shares. A **bar chart** can also be used to show the same sort of information. See figure 64, page 146.

The table in figure 65 is called a **frequency distribution**. It tabulates the frequency with which a mark, weight, salary ... (**class**) occurs. Table 3, page 148, shows the classes grouped. 1–10, 11–20..

The frequency distribution is shown on a **histogram**, with the frequency vertical, the class horizontal. The class is placed in the middle of the block or plotted to the upper limit when the class is grouped. The area of each block is proportional to the frequency. This becomes apparent when the class intervals are uneven, as in table 4. Here the class is plotted against the frequency ÷ class. See table 4 and figure 66(b).

Averages

The **mode** of a frequency distribution is the class with the highest frequency. With grouped classes it is called the modal class. E.g. In table 3 the modal class is the 41–50 mark.

Consider the list 1, 2, 6, 8, 10, 14, 15, 20, 21, 22.

$$\text{The } \textbf{mean} = \frac{\text{sum of the items}}{\text{number of items}}$$

$$= \frac{1+2+6+8+10+14+15+20+21+22}{10} = 11{\cdot}9$$

The **median** of the list arranged in ascending order is the item which bisects the list, or the $\frac{1}{2}(n+1)$th item. In this list it is the $5\frac{1}{2}$ number, i.e. between 10 and 14. The median is 12, their mean. The **lower** and **upper quartiles**, the $\frac{1}{4}(n+1)$th and $\frac{3}{4}(n+1)$th terms respectively, further bisect the lower and upper halves of the list. The lower quartile = 6. The upper quartile = 20.

$$\text{The mean of a frequency distribution} = \frac{\text{sum of class} \times \text{frequency}}{\text{sum of frequencies}}.$$

E.g. mark	0	1	2	3	4	5
frequency	3	5	6	4	2	0

$$\text{mean} = \frac{0\times3+1\times5+2\times6+3\times4+4\times2+5\times0}{3+5+6+4+2+0}$$

$$= \frac{37}{20} = 1{\cdot}85 \text{ marks.}$$

When the classes are grouped e.g. 5–8, 9–12, 13–16 ... use $6\frac{1}{2}$, $10\frac{1}{2}$, $14\frac{1}{2}$ ... as the class.

Cumulative frequency

The frequencies of a frequency distribution are accumulated as in table 5, page 151. The **cumulative frequency** can then be plotted vertically against the class horizontally, to the upper limit of the class in each case (see figure 67). From this curve the median and quartiles can be read off. Use $\frac{1}{2}n$, $\frac{1}{4}n$ and $\frac{3}{4}n$ to locate them if the distribution is large, when this curve is particularly useful.

See the examples on pages 152 and 190.

Dispersion

Range = largest minus smallest. **Semi-interquartile range** = $\frac{1}{2}(UQ-LQ)$. The latter measures the dispersion of the numbers in the middle: the range does not. In the list on page 216 the range = $22-1=21$. The semi-interquartile range = $\frac{1}{2}(20-6)=7$.

Probability

Probability can be expressed as the fraction

$$\frac{\text{number of ways a particular event occurs}}{\text{total number of events}}$$

If a die is cast the chance of a 3 coming up is $\frac{1}{6}$. If a coin is tossed, the chance of 'heads' is $\frac{1}{2}$, 0·5 or 50%. If the die is cast 300 times the expectation of a 3 is $\frac{1}{6}\times 300=50$ times.
If you throw a die the chance of a 3 **not** coming up is $\frac{5}{6}$.
The chance of a 3 or not a 3 is $\frac{1}{6}+\frac{5}{6}=1$, a certainty.
If the probabilities of an event and the non-event are p and q then $p+q=1$.
The chance of a club being drawn from a pack of cards is $\frac{13}{52}=\frac{1}{4}$.
The chance of a spade being drawn is also $\frac{1}{4}$.
The chance that a card drawn from a pack is club or spade is $\frac{1}{4}+\frac{1}{4}=\frac{1}{2}$. The sum of the chances is called **total probability**.

If two dice are thrown, the chance that the first is a 3 is $\frac{1}{6}$. The chance that the second is 3 is $\frac{1}{6}$. The chance that both are 3 is $\frac{1}{6}\times\frac{1}{6}=\frac{1}{36}$. This is **compound probability**. Of all the 36 possibilities only one is two 3's.

The chance that only one is a 3 is calculated thus:
The chance of the first being a 3 is $\frac{1}{6}$. The chance of the second being **not** 3 is $\frac{5}{6}$. The chance of only the first being a 3 is $\frac{1}{6}\times\frac{5}{6}$. But the second can be a 3 and the first not. $\therefore$ the chance of only one 3 is $2\times\frac{1}{6}\times\frac{5}{6}=\frac{5}{18}$.

Two cards are drawn from a pack. What is the chance that they are both aces? Drawing one card, the chance of an ace = $\frac{4}{52}$. Assume success, and do not return the card to the pack. Drawing again, the chance of another ace is $\frac{3}{51}$. The chance of two aces is $\frac{4}{52}+\frac{3}{51}=\frac{1}{221}$.

With more complicated problems a tree diagram can be used. This is a systematic way of writing down all the possibilities. See the examples on pages 155–6.

Chapter 14. Trigonometry (1)

In the right-angled triangle ABC in figure 69, page 157, we have:

$$\textbf{sine } A = \frac{\textbf{opposite}}{\textbf{hypotenuse}} \qquad \textbf{cosine } A = \frac{\textbf{adjacent}}{\textbf{hypotenuse}}$$

$$\textbf{tangent } A = \frac{\textbf{opposite}}{\textbf{adjacent}}$$

These ratios are given as 3- or 4-figure decimals in the tables. As the angle increases from 0° to 90° the value of the sine increases from 0 to 1, the cosine decreases from 1 to 0 and the tangent increases from 0 to ∞. For this reason we **add** the differences when using the sine and tangent tables and **subtract** in the cosine table.

To calculate sides and angles, set out the working clearly.
In figure 69(a): (i) If $AC = 8$ cm, angle $A = 27°$; find BC; (ii) if $AB = 5$ cm, $BC = 8$ cm, find angle A; (iii) if $BC = 6$ cm, angle $A = 50°$; find AC.

(i) $\frac{\text{opposite}}{\text{hypotenuse}} = \frac{BC}{8} = \sin 27° = 0{\cdot}4540$. $\frac{BC}{8} = 0{\cdot}4540 \Leftrightarrow$

$$\begin{array}{r} 0{\cdot}4540 \\ 8 \\ \hline 3{\cdot}6320 \\ \hline \end{array}$$

$BC = 8 \times 0{\cdot}4540 \Leftrightarrow BC = 3{\cdot}63$ to 3 significant figures.

(ii) $\frac{\text{opposite}}{\text{adjacent}} = \frac{8}{5} = 1{\cdot}6000 = \tan A \therefore A = 57°59'$ from the tables.

(Notice that $\tan 45° = 1$ so $A > 45°$.)

(iii) $\frac{\text{opposite}}{\text{hypotenuse}} = \frac{6}{AC} = \sin 50° = 0{\cdot}7660$. $\frac{6}{AC} = 0{\cdot}7660 \Leftrightarrow$

$AC = \frac{6}{0{\cdot}7660}$. The reciprocal of $0{\cdot}7660 = 1{\cdot}305$.

$\therefore \quad AC = 6 \times 1{\cdot}305 \Leftrightarrow AC = 7{\cdot}83$ to 3 significant figures.

If you look at the horizon and raise your eyes to a point above, you have described the **angle of elevation** of the point from your eye. If you look at the horizon and lower your eyes to a point, you have found the **angle of depression** of the point. See figure 71, page 159.

The bearing of B from A in figure 71(b) is found by: (i) standing at A; (ii) looking north; (iii) turning clockwise to face B. The angle turned, expressed as a 3-figure number of degrees, is the bearing of

B from A. In this case it is 050°. Notice that NE is 045°, E is 909°, SE is 135°, S is 180°, SW is 225°, W is 270°, NW is 315°. To find the bearing of A from B stand at B and repeat the process.

The sides of a 30°, 60°, 90° triangle are in the ratio $1 : \sqrt{3} : 2$.
The sides of a 45°, 45°, 90° triangle are in the ratio $1 : 1 : \sqrt{2}$.

From figure 72 it can be seen that $\sin 30° = \cos 60° = \frac{1}{2}$, $\tan 60° = \sqrt{3}$, $\tan 30° = \frac{1}{\sqrt{3}}$, $\sin 60° = \cos 30° = \frac{\sqrt{3}}{2}$, $\sin 45° = \cos 45° = \frac{1}{\sqrt{2}}$

These ratios may be more convenient than 3- or 4-figure decimals on occasions. See the example on page 160.

Three dimensions

Study the diagrams showing the angle between a line and a plane and the angle between two planes. In figure 74(a), the angle between BF and plane $ABHE$ is angle FBE. The angle between the planes $CDEH$ and $ABCD$ is angle ADE or angle BCH. If you have to calculate these angles draw a small sketch of the relevant triangle, labelling the known sides and angles. This helps you to see what other information to look for in the main diagram. Two examples are shown on page 162. The second of them includes the calculation of an angle of elevation.

Plan and **elevation** are the names given to the view from above and the views from the side of an object. E.g. The plan of a **right** cone is a circle with a point marked in the centre. The elevation is an isosceles triangle. The word **right** means that the vertex is above the centre of the base.

The earth as a sphere

Let the radius of the earth = R and the circumference = C.
Great circles have radius = R, e.g. **equator**, longitudes.
Small circles have radius $r < R$, e.g. circles of latitude.
Meridian of longitude = half a great circle passing through the north and south poles. They are measured 0° to 180° east or west of Greenwich (0°). If two points have longitudes adding up to 180° they are on the same great circle. E.g. 30° W and 150° E.
Circles of latitude are parallel to the equator. They are measured in degrees north and south of the equator.
The **radius** of a circle of **latitude** $A°$ north or south = $R \cos A°$.
The circumference of a circle of latitude = $C \cos A°$.
1 **nautical mile** = the length of arc which subtends an angle of 1 minute at the centre of the earth. The **circumference of the earth** = 360 × 60 = **21 600 nautical miles**.
Any point on the earth's surface is at the intersection of a circle of latitude and a meridian of longitude. Hence the point P in figure 76 is at latitude 50° N, longitude 35° W.

Find distances and angles from: $\dfrac{\textbf{arc length}}{\textbf{circumference}} = \dfrac{\textbf{angle}}{\textbf{360°}}$

Chapter 15. Trigonometry (2)

For angles of any size we re-define the ratios. In figure 78, page 169, 0° is on the $+x$ axis, + angles are measured **anti-clockwise.** The axes divide the circle radius r into 4 **quadrants.** In the first quadrant, $0° < A < 90°$; second, $90° < A < 180°$; third, $180° < A < 270°$; fourth, $270° < A < 360°$.
Rotate the line OP to the required angle.

$$\sin A = \frac{y \text{ co-ordinate of } P}{r};$$

$$\cos A = \frac{x \text{ co-ordinate of } P}{r};$$

$$\tan A = \frac{y \text{ co-ordinate of } P}{x \text{ co-ordinate of } P}.$$

The sign of the ratio depends on whether the (x, y) co-ordinates are + or − in that quadrant. Assume that r is +. From figure 78:

1st quadrant:

$\sin A = \dfrac{+y}{r}$ is + $\quad \cos A = \dfrac{+x}{r}$ is + $\quad \tan A = \dfrac{+y}{+x}$ is +

2nd quadrant:

$\sin A = \dfrac{+y}{r}$ is + $\quad \cos A = \dfrac{-x}{r}$ is − $\quad \tan A = \dfrac{+y}{-x}$ is −

3rd quadrant:

$\sin A = \dfrac{-y}{r}$ is − $\quad \cos A = \dfrac{-x}{r}$ is − $\quad \tan A = \dfrac{-y}{-x}$ is +

4th quadrant:

$\sin A = \dfrac{-y}{r}$ is − $\quad \cos A = \dfrac{+x}{r}$ is + $\quad \tan A = \dfrac{-y}{+x}$ is −

In the first, all are +, in the second sin is +, in the third tan is +, in the fourth, cos is +. The initials can be remembered by 'All Stations To Crewe' (All, Sin, Tan, Cos).

E.g. Find (i) tan 135°; (ii) sin 230°; (iii) cos 320°.
(i) 135° is second quadrant, tan is −, $\tan 135° = -\tan(180° - 135°) = -\tan 45° = -1$;

(ii) 230° is third quadrant, sin is $-$, $\sin 230° = -\sin(230° - 180°) = -\sin 50° = -0{\cdot}7660$;
(iii) 320° is fourth quadrant, cos is $+$, $\cos 320° = +\cos(360° - 320°) = \cos 40° = 0{\cdot}7660$.

Alternatively draw a small diagram as shown on page 171.

Graphs
See the graphs of sine, cosine and tangent in figure 80, page 171. $y = \sin x°$ or $f(x) = \sin x°$ or $f: x \rightarrow \sin x°$ has a **maximum** value of 1 when $x = 90°$, a **minimum** of -1 when $x = 270°$. The curve is above the x axis between $x = 0°$ and 180° (first and second quadrants).
The curve of $y = \cos x$ is the sine curve translated 90°. $\sin 40° = \cos 50°$, $\sin 70° = \cos 20°$ etc. The maximum is at $x = 0$, $y = 1$; the minimum at $x = 180°$, $y = -1$. The curve lies above the x axis between 0°–90° and between 270°–360° (first and fourth quadrants).
The curve of $y = \tan x$ has no turning points. As x goes $0° \rightarrow 90°$ $\tan x$ goes $0 \rightarrow \infty$. Vertical lines are drawn through 90° and 270°. The curve **tends** towards these lines. The curve lies above the x axis between 0° and 90° and between 180° and 270° (first and third quadrants).

Figure 81(a) compares (i) $2 \sin x$ and (ii) $\sin 2x$. E.g. When $x = 30°$ (i) becomes $2 \sin 30° = 2 \times \frac{1}{2} = 1$; (ii) becomes $\sin 60 = \frac{\sqrt{3}}{2}$.

(i) Double the ratio. (ii) Double the angle.

When asked to find $x°$ such that $\cos x° = -0{\cdot}5$ remember that there are two such angles between 0° and 360°. For negative cosine they are in the second and third quadrants. $x = 180° - 60°$ and $x = 180° + 60°$; $x = 120°$ and 240°.

$\cos^2 A + \sin^2 A = 1$ and $\dfrac{\sin A}{\cos A} = \tan A$ are two identities linking sine, cosine and tangent. They are true for all values of A.

Note that $\cos^2 A = (\cos A)^2$ the square of the ratio.

E.g. Given that $\cos 30° = \frac{\sqrt{3}}{2}$ find the value of $\cos^2 210°$.

$$\cos 210° = -\cos(210° - 180°) = -\cos 30° = -\frac{\sqrt{3}}{2}$$

$$\therefore \quad \cos^2 210° = \left(-\frac{\sqrt{3}}{2}\right)^2 = \frac{3}{4} \quad \text{or} \quad 0{\cdot}75.$$

Sine and cosine rule
These formulae are used in triangles with no right angle. We need three pieces of information to use either of them. In figure 83 if the

information is (i) 2 angles and 1 side, use the **sine rule**:

$$\frac{a}{\sin A} = \frac{b}{\sin B} = \frac{c}{\sin C};$$

(ii) 2 sides and the angle between them, use the **cosine rule**: $a^2 = b^2+c^2-2bc \cos A$: $b^2 = c^2+a^2-2ca \cos B$: $c^2 = a^2+b^2-2ab \cos C$; (iii) 3 sides, use the cosine rule to find an angle.
Calculations using them are shown on pages 174 and 175.
Do not use these formulae if you can use right-angle trigonometry in the triangle, e.g. in an isosceles triangle, equilateral triangle or rhombus. It is easier to use right-angled-triangle trigonometry.

E.g. In figure 83, page 175, $a = 5$ cm, $b = 8$ cm, angle $C = 146°$. Find c.

$c^2 = a^2+b^2-2ab \cos C$ gives $c^2 = 25+64-80 \cos 146°$
$\cos 146 = -\cos 34° \therefore c^2 = 89-80(-\cos 34°) = 89+80 \cos 34°$
$c^2 = 89+80 \times 0{\cdot}8290 = 89+66{\cdot}32$
$c^2 = 155{\cdot}32 \Rightarrow c = 12{\cdot}5$ cm to 3 significant figures.
Note the way of dealing with the obtuse angle.

Chapter 16. Calculus

Differentiation

Differentiation measures the **gradient** of the tangent, or rate of change, at any point on the curve. The symbol for gradient is $\frac{dy}{dx}$.

If $y = x^n$ $\frac{dy}{dx} = nx^{n-1}$ E.g. if $y = x^2$ $\frac{dy}{dx} = 2x$;

if $y = x^3$ $\frac{dy}{dx} = 3x^2$; if $y = 4x$ $\frac{dy}{dx} = 4$;

if $y = 6$ $\frac{dy}{dx} = 0$; if $y = \frac{1}{x^2}$ or x^{-2} $\frac{dy}{dx} = -2x^{-3}$ or $-\frac{2}{x^3}$

If $y = 4x^3$ $\frac{dy}{dx} = 4 \times 3x^2 = 12x^2$.

If $y = 3x^2 + 7x - 4$ $\frac{dy}{dx} = 6x + 7$.

The differential of a sum is the sum of the differentials.

E.g. Find the gradient at the point (4, 0) on the curve $y = x^2-5x+4$.
The gradient is $\frac{dy}{dx} = 2x-5$; when $x = 4$, gradient $= 8-5 = 3$.

This curve is drawn in figure 22, page 65.

Maximum and minimum
These are the names given to the turning points on a curve. In figure 84(b), A and B are **maxima**, C and D are **minima**. $\frac{dy}{dx} = 0$ at these points.

To determine which is which, substitute values of x just $<$ and $>$ than those of the turning points.

E.g. If $y = x^2 - 5x + 4 \; \frac{dy}{dx} = 2x - 5 = 0$ for maximum and minimum.

$\therefore \quad 2x = 5 \Rightarrow x = 2{\cdot}5$ so $x = 2{\cdot}5$ is a turning point.

If x just $< 2{\cdot}5 \quad \frac{dy}{dx} = -$ If x just $> 2{\cdot}5 \quad \frac{dy}{dx} = +$

The slope goes from $-$ to $+$ $\therefore$ $x = 2{\cdot}5$ is a minimum. $y = -2{\cdot}25$ at this point. The minimum point is $(2{\cdot}5, -2{\cdot}25)$. See figure 22(a).

Velocity and acceleration
If s is the distance from O in time t and $s = t + 3t^2 - 2t^3$ then $\frac{ds}{dt}$ gives **velocity** v and $\frac{dv}{dt}$ gives **acceleration**, continuing the principle of rate of change.

$\frac{ds}{dt} = 1 + 6t - 6t^2 = \text{velocity } v \quad \therefore \frac{dv}{dt} = 6 - 12t = \text{acceleration.}$

Integration
This is 'anti-differentiation'. Given the gradient, find the curve.

If $\frac{dy}{dx} = x^n \quad y = \frac{x^{n+1}}{n+1} + c \quad \left(\text{a constant disappears on finding } \frac{dy}{dx}\right)$.

E.g. if $\frac{dy}{dx} = 2x \quad y = \frac{2x^{1+1}}{1+1} = x^2 + c;$

if $\frac{dy}{dx} = 2x^3 \quad y = \frac{2x^4}{4}$ or $\frac{x^4}{2} + c;$ if $\frac{dy}{dx} = 3 \quad y = 3x + c;$

if $\frac{dy}{dx} = 3x^2 + 4x + 8 \quad y = x^3 + 2x^2 + 8x + c.$

The integral of a sum is the sum of the integrals.

E.g. The gradient at any point on a curve is $6x^2 + 4x$. The curve passes through the point (1, 3). Find the equation.

$\frac{dy}{dx} = 6x^2 + 4x \quad \therefore \quad y = 2x^3 + 2x^2 + c.$ It passes through (1, 3).

$\therefore \quad 3 = 2 + 2 + c \Leftrightarrow c = -1$

and the curve is $y = 2x^3 + 2x^2 - 1$.

The area under a curve

To find the approximate area, use the trapezium rule; to find the **exact** area use integration as follows:

To find the area enclosed between the x axis, the curve $y = x^2 - 3x + 2$ between $x = 2$ and $x = 1$, as in figure 17(a) page 54.

Integrate $\left[\frac{x^3}{3} - \frac{3x^2}{2} + 2x + c\right]_1^2$. Enclose in brackets [ ] quoting limits, substitute limits and subtract:

$(\frac{8}{3} - \frac{12}{2} + 4 + c) - (\frac{1}{3} - \frac{3}{2} + 2 + c)$

$= \frac{8}{3} - 6 + 4 + c - \frac{1}{3} + \frac{3}{2} - 2 - c = \frac{7}{3} - \frac{5}{2} = -\frac{1}{6}$ units^2

A negative area is situated below the x axis.

$\frac{dv}{dt}$ = acceleration $\therefore$ **integrate acceleration** to find **velocity**.

$\frac{ds}{dt}$ = velocity $\therefore$ **integrate velocity** to find **distance**.

E.g. On page 60 the trapezium rule is used to find the approximate area under the velocity/time graph $v = 1 + 6t^2 - t^3$. Here we use integration: $v = \frac{ds}{dt} = 1 + 6t^2 - t^3$

$\therefore \quad s = t + 2t^3 - \frac{t^4}{4} + c.$ This is the distance formula.

The limits are $t = 0$ and $t = 5$ $\left[t + 2t^3 - \frac{t^4}{4} + x\right]_0^5$

$= (5 + 250 - \frac{625}{4} + c) - (0 + c) = 255 - 156{\cdot}25 = 98{\cdot}75.$

I.e. the area under the curve (distance travelled) = 98·75 m.